Recommendations for 'Why Bother?'

Why bother doing anything if it leads to nothing? Stop squandering time, energy, and money on flashy programmes that sparkle one day and are gone the next. This book is a veritable checklist of practical MUST HAVES to ensure lasting and meaningful change to anything worth doing!

Brenton Harder
Head of Enterprise Automation,
FinServ

For anyone who is involved with setting up a new continuous improvement program or looking to bring fresh ideas to an existing program, this book is essential reading. The book combines invaluable practical advice and the latest academic research on the importance of continuous improvement assessments. The book highlights how influential continuous assessments are for all businesses and provides real industry examples of how assessments can be used to initiate and, just as importantly, sustain a culture of continuous improvement.

Dr. Andrew Lahy
Global Head of Strategy and Innovation,
Panalpina

Dr Shigeo Shingo, who was considered one of the world's leading experts on the Toyota production system and lean principles, once responded after being repeatedly asked the question "How can I get senior executives to support our improvement efforts" or "How can I get senior executives to buy into our improvement efforts." The first five times, Shingo told them to "get some good results and show the results to the executives. Then they'll buy in." Finally after the 6th time of getting the same question, Shingo, exasperated, "I can't help it if they do not want to make money!" Developing maturity in behaviors aligned to lean principles is not something to do in your organization because you want to win a prize or because "everyone else is doing it." It is something you do because there is a direct correlation between ideal behavioral maturity and ideal business results. Butterworth, Jones, and Hines do a wonderful job of bringing together decades of thought, coaching and experience on the assessment of organizations, teams, and leaders to foster a continuous improvement culture and drive maturity that unlocks business results.

Peter Barnett
Master Coach, Master Blackbelt
Shingo Examiner Board Member

The approach to driving continuous improvement is simple and clear in Why Bother, which describes how to define and assess your organisation's unique culture. Numerous case studies reveal ways to best measure and drive key performance behaviours, which detail how the processes work and the benefits that are created.

Cheryl Jekiel,
Founder, Lean Leadership Center, Inc.

Understanding how to shape our culture through behaviors formation and KBI was a great eye-opener for us in ST. In this book, Chris Butterworth, Morgan Jones and Peter Hines are revealing not only the Why and What but also the How to do take this journey. A must read.

Olivier Ardesi,
Lean and Change Office Director, SD Microelectronics

Embedding and sustaining a culture of Continuous Improvement is challenging. Having spent years training, coaching and assessing tools, frameworks and projects, we realised something was missing. We found that defining, managing and assessing behaviours were missing links to make culture change stick. This easy-to-read book is a fantastic how-to guide, full of real-life examples. I look forward to our next transformation journey, armed with this invaluable roadmap to behavioural deployment.

Indrajit Ray
Senior Manager Customer-led Transformation, NBNC

The authors have provided not only a very practical how to and what to do in setting up an assessment program for measuring the Continuous Improvement behaviours but the why. The case studies bring out the underlying behaviours to life. A must read for anyone wanting sustainability of their CI transformation!

Graham Cockerton
Industry Lead—Global Business Solutions, IBM

This is a valuable reference guide for anyone undertaking a lean, cultural improvement transformation. The illumination of how to embed the desired behaviours to sustain the improvement journey is insightful and the case studies instructive. It provides a valuable roadmap that I will continue to reference well into the future.

Fionna Millikan
Head of Development and Planning, BHP

Why Bother?

Why and How to Access Your Continuous-Improvement Culture

Chris Butterworth, Morgan Jones, Peter Hines

A PRODUCTIVITY PRESS BOOK

First published 2022
by Routledge
600 Broken Sound Parkway #300, Boca Raton FL, 33487

and by Routledge
2 Park Square, Milton Park, Abingdon, Oxon OX14 4RN

Routledge is an imprint of the Taylor & Francis Group, an informa business

© 2022 Chris Butterworth, Morgan Jones, Peter Hines

The right of Chris Butterworth, Morgan Jones, Peter Hines to be identified as authors of this work has been asserted by them in accordance with sections 77 and 78 of the Copyright, Designs and Patents Act 1988.

All rights reserved. No part of this book may be reprinted or reproduced or utilised in any form or by any electronic, mechanical, or other means, now known or hereafter invented, including photocopying and recording, or in any information storage or retrieval system, without permission in writing from the publishers.

Trademark notice: Product or corporate names may be trademarks or registered trademarks, and are used only for identification and explanation without intent to infringe.

Library of Congress Cataloging-in-Publication Data
A catalog record for this title has been requested

ISBN: 9781032028293 (hbk)
ISBN: 9781032028286 (pbk)
ISBN: 9781003185390 (ebk)

DOI: 10.4324/9781003185390

Typeset in Garamond
by Newgen Publishing UK

We are proud to support Black Dog Institute's work to improve the lives of people impacted by mental illness and suicide. One hundred percent of the authors' royalties goes toward Black Dog Institute's research and education programs to help raise awareness and create real-world solutions. One in five of us will experience symptoms of mental illness in any given year, and approximately sixty percent of these people will not seek help. As the only medical research institute in Australia investigating the importance of maintaining our mental health throughout our lifetime, Black Dog Institute aims to create a mentally healthier world for everyone. Black Dog Institute works to improve the lives of people impacted by mental illness through their pioneering research, high quality clinical services, and national education programs.

These royalties will be directed toward education and awareness activities and supporting innovative research into new clinical solutions for all Australians. You are helping Black Dog Institute raise vital funds for:

- researching suicide prevention and clinical treatment;
- expanding our education programs to reach more communities, schools, and health professionals; and
- developing apps and websites to serve as real-time mental health tools so that people can manage depression, anxiety, and stress at their own pace.

Contents

List of Figures xiii

List of Tables xvii

Foreword xix

Acknowledgments xxi

About the Authors xxiii

    Introduction 1

1   **Why Bother Getting Sponsorship?** 9
Having assessments not only seen as important by the senior leadership but also owned by one of the executives. Sponsorship supports the gathering of evidence to assess whether the desired behaviors that you want to sustain in the organization are starting to embed and allow you to focus your efforts on areas that need more attention.

2   **Why Bother Defining Behaviors and Linking Systems and Behaviors?** 15
Having clarity on what the desired behaviors are, deploying and then managing them is fundamental in sustaining a CI cultural transformation. Having specific measures of these behaviors in the form of KBIs drives the continual focus of refining the business systems and embedding these behaviors into the organizational DNA.

3   **Why Bother Assessing and Managing Behaviors?** 33
Continuously maturing behaviors for existing employees and onboarding new employees into the desired behaviors is essential to sustaining a CI culture. There is a wide range of activities

ix

needed to manage and embed ideal behaviors. A behaviorally based maturity assessment is key to enabling us to track progress and identify actions needed to correct slippage or progress to the next level. Assessments that only review tools or systems will not support the development nor sustain the CI culture of the organization.

4 Why Bother Designing Your Own Strategic Level Behavioral Assessment System? **71**
Each organization has a unique context with a different start point, different culture, different business language, geographic complexities, and organization structure. As such, a lot of thought has been given to the approach, design, and content of the assessment and customization to local requirements is critical to success. Designing your own assessment system will support the long-term sustainability of your CI culture. This chapter explains a detailed step-by-step guide showing how to design and build your own CI maturity assessment system.

5 Why Bother Defining Behaviors and KBIs? **103**
Key Behavioral indicators are essential to developing and maturing ideal behaviors that enable any organization to embed a sustainable culture of CI. They can be used across the organization in any area or function. This in-depth case study by Professor Peter Hines and STMicroelectronics explains why KBIs are critical and illustrates their application in the HR function.

6 Why Bother Focusing on the Type of Conversations People Have? **137**
It is clear from Kevin's chapter that how we have conversations is far more powerful than the words used. Indeed, the same words can be used to "tell, suggest, or ask," depending entirely on what "voice" we use. It also clearly shows how the types of conversation and voices used change significantly as a CI culture matures. As such, any behavioral assessment system needs to be able to assess how people are using their voices and if they are applying the appropriate dialogue in a skillful way.

7 Why Bother Seeing Where It Has Worked and the Lesson Learned? **159**
While every organization has unique context and requirements there is nevertheless a lot of value to be had from understanding

the approaches that other organizations have taken. In this wide range of cases, the authors share in their own words what they did. How it was applied, and the lessons learned. The case studies are from several different sectors and include telecommunications, logistics and distribution, aircraft maintenance, car manufacture, financial services, mining, and food production. The final case explores how maturity assessments have been adapted and undertaken using a virtual approach necessitated by COVID-19 restrictions.

Case Studies:
NBN Co. Case Study by Indrajit Ray, Clyde Livingston, and Richard Perry..159
Panalpina Case Study by Andrew Lahy, Maria Pia Caraccia, and Mike Wilson 167
Airbus Australia Pacific Case Study by Kim Gallant 177
CBA Case Study by Morgan Jones..185
Bakkavor Desserts— Measuring the Maturity of the CI System Case Study by John Bowman and Leighton Williams..................198
Virtual Assessments Case Study by Morgan Jones................212

8 **Why Bother Aligning Assessments, Assessors, and Calibration?**..**221**
It is important to have consistency in the assessments to show they are both robust and credible to ensure the recommendations will drive the best focus for action planning. While there is always going to be a certain level of subjectivity it is important to make the assessments as consistent as possible to avoid potential confusion and mixed messages. Each assessor must be trained in-depth to a standard that can be universally applied but at the same time recognizes their expertise and experience. The customer experience must be consistent.

9 **Why Bother Having a High-Level Roadmap to Deploy?**...........**245**
A high-level visual roadmap is an invaluable planning and communication tool. It is a quick way to show the organization what is involved. It enables the creation of detailed action plans and milestone planning and provides a framework for initial design and ongoing improvement of the assessment system. Each organization has a unique context with different a starting points, different cultures, different business languages,

geographic complexities, and organization structure. As such the roadmap in this chapter is meant as illustrative and we encourage people to adapt it to their own specific context and requirements.

10 Why Bother Doing Assessments?...**251**
The CI maturity assessment needs to be integrated into the strategic business planning cycle so that outputs and opportunities can be incorporated into forward planning. They should not be a stand-alone activity to business planning but rather a key check on progress and a key input for consideration in action planning. This chapter gives a summary of the key bullet points for the why, the what, and the how of developing and undertaking CI maturity assessments.

Appendix **255**
Index **259**

Figures

Figure 0.1	The PDCA Cycle	5
Figure 2.1	Ideal Results Require Ideal Behaviors	16
Figure 2.2	KBIs, Processes, and KPIs	17
Figure 2.3	Systems and Behaviors	21
Figure 2.4	The Learning and Development System	23
Figure 2.5	Behavioral Formation and Deployment	23
Figure 2.6	The Shingo Model	26
Figure 2.7	The Shingo Dimensions and Principles	27
Figure 3.1	A 5S Audit Extract Example	34
Figure 3.2	Six-Step Approach to Gemba Walks	46
Figure 3.3	Learning and Development System Purpose Statement	50
Figure 3.4	Impact of Behavior-Focused Recruitment	59
Figure 3.5	Expected Behaviors	61
Figure 3.6	Behavioral Recruitment Assessment	62
Figure 3.7	Behavioral Assessment Form Example	64
Figure 3.8	Extract from Behavioral Playbook	66
Figure 3.9	Extract from Leader Playbook	67
Figure 4.1	Example Simple Summary Assessment	79
Figure 4.2	Summary of Phases of Maturity	84
Figure 4.3	Example Statements at Maturity Levels for a Single Element	85
Figure 4.4	Plan, Do, Check, Act Cycle for Undertaking an Assessment	98
Figure 5.1	Tools, Systems, and Principles	104
Figure 5.2	Enterprise Excellence and the Disconnected Bridge	105
Figure 5.3	Core Operating Systems	106
Figure 5.4	KPIs and KBIs	108
Figure 5.5	Behavioral Formation and Deployment	110

Figure 5.6	MAJU Management Practice Journey	114
Figure 5.7	Developing the MAJU Management Practices	115
Figure 5.8	Head of Department Experiment with the Management Practices	115
Figure 5.9	Support Kit Card Set for Step Back Management Practice	116
Figure 5.10	Support Kit for Empowerment Management Practice	117
Figure 5.11	Support Kit for Sustainability Management Practice	118
Figure 5.12	MAJU Corner	119
Figure 5.13	Lean and Play Process	120
Figure 5.14	The Eight Routines (or Behaviors) that were Initially Defined	120
Figure 5.15	The Three Prioritize Routines and Three Management Practices	120
Figure 5.16	KBI Cards	121
Figure 5.17	Lean & Play and KBI Process	121
Figure 5.18	KBI Template	122
Figure 5.19	KBI Maturity Assessment	123
Figure 5.20	KBI Maturity Matrix	123
Figure 5.21	KBI Process	125
Figure 5.22	Visual Management of KBIs in HR	127
Figure 5.23	Visual Management of Process and Continuous Improvement in HR	128
Figure 5.24	Visual Management of KPIs in HR	129
Figure 5.25	Example of KBI Deployment in IT	130
Figure 5.26	KBI Behavior Enhancement Survey, STMicroelectronics, Noida	131
Figure 5.27	Level 3 KBIs Framework	134
Figure 5.28	The New World Kirkpatrick Model	135
Figure 6.1	The Journey to a Sustainable CI Culture	139
Figure 6.2	SoundWave Model	142
Figure 6.3	High Variation Phase	143
Figure 6.4	Typical SoundWave Profile at High Variation Phase	144
Figure 6.5	The Continuum of Conversation	145
Figure 6.6	Phase 2 Reducing Variation	146
Figure 6.7	SoundWave Profile at Phase 2	147
Figure 6.8	Conformance and Standards	150
Figure 6.9	SoundWave Profile at Phase 3	150
Figure 6.10	Continuous or Generative Improvement	152

List of figures ■ xv

Figure 6.11	SoundWave Profile in Phase 4	153
Figure 6.12	The SoundWave Model	157
Figure 7.1.1	Excerpt from NBN Co. Process Maturity Playbook	160
Figure 7.1.2	Summary Descriptors for Process Maturity Levels	161
Figure 7.1.3	Process Hierarchy	162
Figure 7.1.4	Assessor Checklist	164
Figure 7.1.5	Sample Report Extract	165
Figure 7.2.1	LogEx Pyramid	170
Figure 7.2.2	The Lean Business Model© S A Partners	172
Figure 7.2.3	The Transition from the LogEx Pyramid, to the Lean Business Model, to the Shingo Enterprise Excellence Model	174
Figure 7.3.1	The Target of the Airbus Australia Pacific Transformation	179
Figure 7.3.2	Continuous Improvement Principles at Airbus Australia Pacific	180
Figure 7.3.3	Applying a Measurement System to Ensure Progress Against Strategy	180
Figure 7.3.4	The SA Partners Lean Business Model	182
Figure 7.3.5	Extract from the Lean Maturity Procedure	183
Figure 7.4.1	The 3Cs Productivity Program	186
Figure 7.4.2	The CBA Accreditation System	187
Figure 7.4.3	The CBA Habits Working as a System	189
Figure 7.5.1	The Assessment Model	199
Figure 7.5.2	Program Design and Deployment	200
Figure 7.5.3	Maturity Levels	202
Figure 7.5.4	System Elements	203
Figure 7.5.5	Assessment Scoring Summary	204
Figure 7.5.6	Introduction to the Feedback	206
Figure 7.5.7	Example Findings	207
Figure 7.5.8	Quotes and Observations	208
Figure 7.5.9	Status Tracker	209
Figure 7.5.10	Assessment Recommendations	210
Figure 7.5.11	Summary Action Plan	211
Figure 8.1	An Example Assessment Maturity Roadmap	223
Figure 8.2	Example Maturity Assessment Process	224
Figure 8.3	Example CI Behaviors Across Different Levels of the Organization	235

Figure 8.4	Example Measurement Scale for Team Members on CI	238
Figure 8.5	Descriptors to Help Score Definitions	239
Figure 8.6	Potential Assessor Learning Journey	242
Figure 9.1	The Assessment System Development Roadmap	246

Tables

Table 0.1	Audits Versus Assessments	4
Table 2.1	Create Value for the Customer Example Behaviors	28
Table 2.2	Constancy of Purpose Example Behaviors	29
Table 2.3	Think Systemically Example Behaviors	30
Table 3.1	Gemba Walk Themes	45
Table 4.1	Behavioral Assessment Scale	88
Table 5.1	Lag KPIs, Lead KPIs, and KBIs	108
Table 5.2	KBI Survey Results in HR, STMicroelectronics, Noida	132
Table 6.1	Summary of Observed Behaviors at Different Maturity Levels	141
Table 6.2	Summary of Behavior Needing Conversation	142
Table 6.3	Risks, Challenges, and Counter Measures (Phase 2)	148
Table 6.4	Risks, Challenges, and Counter Measures (Phase 3)	151
Table 6.5	Risks, Challenges, and Counter Measures (Phase 4)	154
Table 7.4.1	Maturity Levels of the CBA Accreditation Program	188

Foreword

"Why Bother?"

The title of this book reminds a lot of questions I have asked frequently—Why is there a Shingo Prize? Both questions are based on using assessments as a way to drive improvements. I can honestly say that the Shingo Institute would not exist if assessments did not bring value to the organizations being assessed.

As the Executive Director of the Shingo Institute, home of the Shingo Prize, I've had the opportunity to participate in some way or another in hundreds of assessments. Of course, this includes the dozens of assessments performed by our Shingo examiner teams each year. But it also includes the assessments performed during workshops, and opportunities I have to assess an organization while visiting for a study tour or showcase event. On average, It's estimated that I visit 50–60 of the best organizations in the world every year. I am blessed to see greater organizations doing greater work than perhaps anyone else in the world.

So, "Why Bother?" with assessments? Let me share what common results were reported by organizations applying for the Shingo Prize:

- A safer workplace: Fewer accidents: Fewer days lost due to workplace injuries;
- Higher morale: Decreased turnover;
- Better quality: Fewer defects: Improved yields: Fewer customer complaints: Improvements in product quality;
- Better productivity: Decreased costs: Improved process efficiencies;
- Shorter lead times: Improved on-time deliveries; and
- Increased sales: Customers appreciate better quality and shorter lead times so they buy more.

Through the examples cited in this book, Butterworth, Jones, and Hines demonstrate that effective assessments can help organizations identify gaps. And by closing those gaps, real, measurable improvement is achievable.

I'm often asked is it better to have assessments done by internal or external people. The answer I give to that question is "Yes!" Internal people generally know the business better and might see gaps specific to the business or the industry. Internal assessors can also bring unique opportunities for knowledge-sharing. But it never hurts to have some expert outside eyes come and observe too. Outside eyes often see the waste that internal people might take for granted. The results commonly reported are usually achieved through the effective use of both internal or external assessments.

The focus placed by Butterworth, Jones, and Hines on assessing the *behaviors* that drive results is the key to being able to achieve outstanding results. Ideal results require ideal behaviors. The very best organizations focus on behaviors that drive ideal results. For example, fewer workplace accidents don't just happen because someone posts a "Safety First" sign. Fewer workplace accidents happen because everyone is dedicated to identifying potentially unsafe situations, and reporting those situations (a behavior), and then putting in error-proofing devices to prevent the potentially unsafe situation from ever causing a safety accident (another behavior). Behaviors like these can be observed and measured in an assessment process.

Perhaps one of the best contributions by Butterworth, Jones, and Hines in this book is the case studies. Each case study demonstrates how each assessment performed can serve as a road map for the next phase of improvements.

With all that said, there is no magic recipe for how and when to use assessments. Each organization needs to identify the frequency and scope of their internal assessments. Each organization needs to determine the cadence and focus of inviting in outside eyes. I encourage you to try different approaches to see the value of each. It's okay to learn as you go. You will learn more about your gaps and your approaches with each attempt. The biggest mistake is not trying assessments at all. It's worth the "bother."

Ken Snyder
Executive Director
Shingo Institute
Utah State University
Logan, Utah, USA

Acknowledgments

We would like to thank the thousands of people we have worked with on assessments over many years. We learned something from every conversation and have attempted to share these lessons throughout this book.

In addition, a special thank you to those who have contributed case studies and those who were kind enough to read and give feedback on the draft manuscript. In particular, they include Amelia Deich, Irene Teo, Camille Pied, Kailash Chandra Joshi, Ashish Chawla, Kevin Eyre, Indrajit Ray, Clyde Livingston, Richard Perry, Richard Steel, Andrew Lahy, Maria Pia Caraccia, Mike Wilson, Kim Gallant, Milan Gajic, Leighton Williams, John Bowman, Scott Kean, Stephen Dargan, Sandie Butterworth, Nicole Gallant, Brenton Harder.

About the Authors

Chris Butterworth

Chris is a multi-award-winning author, speaker, and coach. He is a certified Shingo Institute master-level facilitator and a Shingo Institute Faculty Fellow and examiner.

He coaches executive teams and transfers continuous improvement knowledge across all levels of an organization. He has extensive leadership and consulting experience in a wide range of sectors.

Chris is the winner of the Best New Speaker of the Year Award for The Executive Connection (TEC) for his talk on Lean Thinking. He is the co-author of the widely acclaimed Shingo publication award-winning books *4+1: Embedding a Culture of Continuous Improvement* and *The Essence of Excellence* and is also the editor of the Shingo Institute book *Enterprise Alignment and Results*.

chris@cbenterpriseexcellence.com

www.cbenterpriseexcellence.com

About the Authors

Morgan Jones

Morgan has over thirty years of experience in Lean and twenty years in Six Sigma. He is a pragmatic and experienced improvement leader. He has delivered over $2.1 billion in hard savings to organizations and improved customer-staff experiences and health and safety. The sustainability of business improvement has resulted in over twenty-three international awards and he has chaired twenty-seven international conferences around business improvement. Morgan is an international award-winning author, and he has led a business unit with an overall profit and loss accountability of $167 million.

He is also a Chartered Engineer, Certified Master Black Belt, Lean Master, and Executive Coach. Morgan has leadership experience in marine, manufacturing, government, military, mining, utilities, telecommunications, oil and gas, banking, and supply chain.

Peter Hines

Professor Peter Hines is the co-founder of the Lean Enterprise Research Centre (LERC) at Cardiff Business School. LERC has grown to be the largest academic research in Lean globally.

He has undertaken extensive research into Lean and written or co-written twelve books, including the Shingo Prize-winning books *Staying Lean, Creating a Lean & Green Business System*, and *The Essence of Excellence*.

Peter now runs the Enterprise Excellence Network, providing European-based forums with on-site benchmarking, learning, and networking opportunities. He also co-founded the global www.PeopleValueStream.com, which is a not-for-profit portal for research, learning, and dissemination into the people side of the Enterprise Excellence Network. His academic links continue as a Visiting Professor at Waterford Institute of Technology, Ireland, and a Faculty Fellow of the Shingo Institute at Utah State University, USA.

Peter can be contacted at peter@enterpriseexcellencenetwork.com

Introduction

All the authors have been passionate advocates of Continuous Improvement (CI) for over twenty-five years. In our book *4+1: Embedding a culture of continuous improvement in financial services*,[1] Jones and Butterworth sought to explain some of the key habits needed to embed CI tools and mindsets throughout an organization. We also touched briefly on how to assess the extent to which those tools were embedded throughout the various stages of the CI journey. In this book, we aim to expand on the whole concept of assessments from both a top-down and bottom-up perspective and show how they are critical to the long-term sustainability of a CI culture.

"Why Bother?" is deliberately intended to be a thought-provoking title and reflects the authors' experiences of some of the struggles that organizations go through to get real value from an assessment system. One of the most common reactions we see from many people can be summarized as "not another audit from HQ!"

Our purpose in writing this book is to explain just how useful the assessment process can be, how to design an assessment process that will deliver real value, how to engage people in the process, and how to help people explain "why bother" doing assessments.

Our target readers are the people and culture leaders who want to have an overarching understanding of where their organization is at in terms of CI culture and what steps are required to move that culture to the next level. With over one hundred different definitions of "culture" this can be quite a challenge. From the authors' perspective, we define a CI culture as one where every person at every level is striving to make tomorrow better than today for their colleagues, their customers, and their suppliers. CI culture does not have an end point.

We believe that business leaders will also find this book useful because it gives them a framework to check if existing approaches are delivering value and, if not, to design new ones. In many ways, a good assessment system can provide a powerful lead indicator that can guide decision making and help set priorities. Leaders and their executive teams often find it difficult to measure progress in CI culture and return on investment for the effort and cost involved (other than often inflated and/or short-term financial numbers). An appropriately designed and executed assessment system will provide real progress data and guide what actions need to be taken to keep on track and progress further.

CI leaders often find themselves talking a language different from their colleagues and often resort to visible artefacts or tools to demonstrate progress. For example, we have X percent of our teams using a visual management board. In reality, measures like this do not tell us how effective our CI culture is, nor whether it has any chance of being sustained. We hope to illustrate with this book how to create a common language to assist CI leaders and everyone across an organization understand the current level of CI culture within their company and what they can do next to realize even more improvements.

The book is intended to be a practical how-to handbook with many real-life examples and case studies not only of what works well but also of what does not work and the key things to avoid. We aim to provide details on an approach that gives sufficient information to the reader to be able to design and implement their own assessment system. While we will cover what a good assessment system looks like and how to use one, critically what we will also explore is why bother to do them at all.

We do not advocate a one-size-fits-all, off-the-shelf answer, but provide readers with the templates, tools, and understanding to create an assessment system for themselves within a guiding framework.

It is our hope that *Why Bother* will become a guidebook that will be used in many different organizations around the world and that you will find it as enjoyable to read as we did to write.

0.1 What Is In a Name?

One of the big challenges in suggesting an assessment is that, for many people, it is immediately perceived as yet another audit—another thing that is going to suck up valuable time, create jobs for head office, and be a complete waste of time. Unfortunately, the sad thing is that this is often true. The starting point when thinking about designing an assessment system is to be very clear on the purpose.

So, let's state right off what it isn't.

0.2 It Is Not an Audit

A common dictionary definition of an *audit* is "to make an official examination of the accounts of a business and produce a report to officially examine the financial accounts of a company"[2] and it is often seen as a negative thing. We see organizations preparing for an audit because they do not want to "fail" or "get caught out." Some organizations spend weeks preparing everyone for the annual audit and auditors are often viewed as someone to hide things from—"don't tell the auditors" is whispered in many coffee shop conversations and often becomes a light-hearted (but revealing) refrain to anything that goes wrong.

While many would argue this perception is incorrect, it remains a prevailing attitude in many organizations. The CI community has a lot to answer for here and we would be the first to say that we have hopefully learned a lot from our past mistakes. There is a need to justify the jobs, the training, and the investment. Leaders want to know how many people have been trained in what tools; for example, how many managers are Six Sigma Black Belts or how much money have consultants saved us? The CI community has embraced audits as a tool, for example, 5S audits, visual management board audits, and so on. The list has become endless and often ever more complex in a mistaken attempt to make the measurement less subjective or more rigorous.

In one example, a large multinational changed its simple progress check to a three hundred-question audit that required hours to complete and lost all value as it deteriorated into a tick-box exercise. To use another example, one improvement program in a large, multi-site retail business was measured entirely on the budget savings made to the operating cost of each store. The program reported massive success with double-figure savings for two consecutive years. The CI leader was promoted and everyone was pleased with the financial results. Unfortunately, in year three things started to fall apart—quite literally. The biggest savings had come from slashing almost all the preventative maintenance and running the equipment until it burned out. Year three required massive investment in infrastructure, new equipment, and disruption to customers and internal teams, which was a direct result of the cost "improvements" reported.

Quite rightly, finance vice presidents and budget holders want to know if they are getting value for money and a good return on investment. This is good business practice. The CI community's response to try to justify their existence is understandable. Their concern, though, is that a quantitative approach to measuring CI drives undesired behavior and limits the chances of sustainability. Instead, we need to rethink what we are measuring and why we are measuring it.

One of the main reasons many assessment systems measure the wrong thing is a lack of clarity around the purpose of an assessment.

So, what is the purpose of an assessment? This is a valuable discussion to have and each organization will want to include their own perspective and context, but at a high level we believe the purpose should be as below:

> **Purpose:** Learn where we are now and what the key steps are that we need to take to get even better.

Table 0.1 below shows some of the more common issues associated with a traditional audit approach and how we would prefer people to see assessments.

To try to overcome some of these issues, we have found it necessary to explain to people that the check on progress is an assessment and not an

Table 0.1 Audits Versus Assessments

I will be in trouble for any mistakes they find.	We need to understand what is not working and support your ideas on how to fix it.
They are trying to catch me out.	We need to understand what is not working and support your ideas on how to fix it.
Head office time waste of time.	This is a valuable exercise that will give me real insights and tangible actions.
CI team are just trying to justify their existence.	This is about how the CI team can help me to be more successful.
CI team just want to take the credit for all our hard work.	This will help the organization to recognize our efforts and share best practice.
Finance wants to know how much they can cut the budgets by for next quarter.	We need to understand what is working well and what needs further support so we can target investment.

audit. This is often still met with scepticism and always will be until people see how the results are used. It is not what we do that is important but how we do it and how we use the results.

One of the key approaches and important mindsets to have in improvement at any level is Plan, Do, Check, Act (PDCA). This is often used by a team to structure action plans or improvement projects and is represented in a simple graphic, such as Figure 0.1 below:

A very high-level summary of PDCA is given below. It is not only something that applies to projects but also is a mindset and a way of working.

- **Planning** (P)—involves understanding what the customer values, being clear on the desired outcomes and setting out the key milestones and actions required.
- **Doing** (D)—involves undertaking the elements in the plan.
- **Checking** (C)—involves reviewing if the actions taken in D have been completed and if they have delivered the expected outcome.
- **Acting** (A)—involves incorporating lessons learned from C and conducting either a wider rollout or redesign that necessitates going back to P.

In reality, any significant program of work will have multiple PDCA cycles and will often have sub-level PDCA cycles within the higher PDCA steps. This is illustrated for the assessment system in Chapter 9.

Figure 0.1 The PDCA Cycle.

Unfortunately, all too often we see a lot of Plan and Do in a repeating cycle with insufficient Check. This has numerous risks associated with it.

How can the organization know if what it is doing is effective without a structured review of the progress being made and the results being delivered? Too often the program delivery becomes the goal and the measures are used to track the implementation of the tools; for example, the percentage of teams that have a visual management board or the number of people trained in Lean Six Sigma. These tell the organization little about how effective their improvement initiative is or what they need to do better.

Without the effectiveness check, the risk is that the organization will continue to do the same thing and waste time and effort replicating things that are not delivering the required result. How will the organization understand which activities it should continue doing more of and which need a different approach?

Some of the other benefits of a well-designed assessment system include:

Benchmarking best practice—an assessment should help us to learn from what has worked well and replicate it.
How do we know if the results we are seeing are deliberate consequences of the actions we have taken or are just lucky?
Are we getting the benefits we expected?
What are we learning about our approach and what do we need to build on or change?
How do we know if the progress made will be sustained?

Without a structured Check in place, we risk not being able to answer any of these questions.

In many cases, it is likely that financial benefits are being tracked (especially true if external consultants are being measured against this) but these alone are not enough. What is needed is a much deeper *check* of how we are doing and, most importantly, what we need to do next.

However, what is most important is not the results themselves, but rather how we achieve them. As such, the assessment system needs to assess and review the behaviors in the organization. This will be explored in more detail throughout the book.

Notes

1 Morgan Jones, Chris Butterworth, and Brenton Harder, *4+1: Embedding a Culture of Continuous Improvement in Financial Services* (2nd edn.) (Melbourne: Action New Thinking Ltd, 2019).
2 Cambridge Dictionary, Cambridge University Press.

Chapter 1

Why Bother Getting Sponsorship?

Chapter Summary

Having assessments not only seen as important by the senior leadership but also owned by one of the executives. Sponsorship supports the gathering of evidence to assess whether the desired behaviors that you want to sustain in the organization are starting to embed and allow you to focus your efforts on areas that need more attention.

Having a strategic behavioral indicator (KBI) that the Sponsor can show alongside the outcome KPIs to ensure the benefits will be sustainable.

For a Lean transformation to succeed it needs to employ the services of a "Champion" and the support of a "Sponsor"; although, depending on the scope of deployment, the roles of the Sponsor and Champion could be rolled into one person. The Sponsor is the transformation advocate in the leadership team who "sponsors" the transformation, while the Champion has a substantial stake in the outcome and "owns" the transformation of their business unit or department. The project Champion is usually a senior leader or one who heads a certain functional area and therefore has formal authority and ownership of the process being improved.

Champions are mainly responsible for the deployment of Lean transformation within their organization. They identify, propose, and assess potential projects that are aligned with the goals of the organization.

They also set priorities and lead project planning, strategizing, and implementation to maximize benefits and ensure project success.

In today's business world, what is expected of a Sponsor's role has moved away from being an expert with access to vast amounts of knowledge and data. The Sponsor's role now is to create three new things for the transformation: business context, business or department relevance, and individual meaning. Sponsors then help the team to capture these new things succinctly and package and deliver them in a way that is truly understood. Anyone can come up with great improvement ideas—usually Sponsors and Champions—but others can also come up with the same ideas. It is the person who can turn those ideas into something beneficial to the organization that makes the difference.

For large-scale projects, there may be several Champions but there can be only one Sponsor. For smaller projects, however, one person can perform the roles of both Sponsor and Champion. For brevity and simplicity, the word "Sponsor" is used throughout this book for both roles.

To drive a successful Lean transformation, the Sponsor in the leadership team needs to continually advocate and contextualize the transformation. Determining a way of providing evidence of progress and being driven by data and evidence is key to attaining a successful transformation. The Sponsor advocates behavior change based on the evidence from the assessments of the development and embeds these behaviors.

The Sponsor owns the overall transformation and relies heavily on the assessments. The Sponsor defines the goals and assesses the eventual success of the transformation. The project Sponsor also champions and advocates for the behaviors to be adopted by the other members of the executive management team within the business.

1.1 Why Is Having the Right Sponsor Important?

Now that we understand why having a Sponsor is important, who is the right person to be the Sponsor? First, the best sponsor is someone who is respected on the leadership team. Second, they need to be persistent in driving change.

There are four key characteristics of a good sponsor—Clarity, Criticality, Commitment, and Consistency.

Clarity—Provides transformation goals, performance standards, and expectations. Procedures, roles, and responsibilities should be compellingly clear to increase understanding and acceptance of the transformation imperatives. It is important to measure progress through the transformation to maintain momentum.

Effective leaders don't have to be passionate. They don't have to be charming. They don't have to be brilliant ... They don't have to be great speakers. What they must be is clear. Above all else, they must never forget the truth that of all the human universals—our need for security, for community, for clarity, for authority, and for respect—our need for clarity ... is the most likely to engender in us confidence, persistence, resilience, and creativity.[1]

Perhaps, the most potent argument for clarity is that if we are not clear about an opportunity then we cannot take advantage of it. If we do not understand a plan, we cannot implement it. When we are uncertain of what we are trying to achieve, we cannot get there.

Criticality—A sense of urgency should be established. The transformation should link strategy to an organization's vision. This can be done by aligning business issues, priorities, processes, and metrics. Criticality helps eradicate complacency and engages critical attention to project imperatives.

A false sense of urgency is pervasive and insidious because people mistake activity for productivity ... a sense of urgency is not an attitude that I must have the project team meeting today but that the meeting must accomplish something important today.[2]

The dictionary defines *criticality* as "the quality, state, or degree of being of the highest importance."[3] In business operations or in financial terms, criticality describes the ranking of the severity of the various ways in which a system, device, or process can fail, the frequency of failure, and the consequences of their failure. The purpose of this ranking is to guide organizations to choose which battles to wage first.

Commitment—Nurturing commitment creates passion for success, intensifies dedication and focus, builds both personal capability and team effectiveness, sustains energy and motivation, and builds the infrastructure of accountability.

The harder you work, the harder it is to surrender.[4]

Commitment, as the dictionary would describe it, is "the trait of sincere and steadfast fixity of purpose or that state of being bound emotionally or intellectually to a course of action or to another person or persons."[5] Commitment to a project (including its vision, objectives, activities, and requirements) would then mean belief and acceptance of the project, translated to a willingness to exert considerable effort to achieve purposes and objectives, and a desire to keep working for the group until completion.

Consistency—Observing consistency in the tangible and intangible aspects of project management builds trust, fosters constancy to purpose, strengthens commitment, and dispels unpredictability and myths.

"Make your mold. The best flux in the world will not make a usable shape unless you have a mold to pour it in."[6]

You just had a great kick-off meeting with your transformation team, a signal that the project is off to a great start. Everyone was excited about the project. Each team member was clear about what needs to be done, how to do it, and when to do it. After a few days/weeks, you take out your checklist and make a cursory evaluation of what has been accomplished so far. To your horror and dismay, not a single box contains a tick mark! No one has touched or moved anything! If you ask why, you will realize your team has stumbled on several roadblocks. If this can happen at the start of a transformation, it can very well happen in the middle and even toward the end.

1.2 What Do You Need for Success?

So how do you set up the Sponsor for success?

Selecting and appointing the Sponsor is relatively easy. Selecting a senior executive who will champion and set them up to be successful and drive the maturity assessment program is a little harder.

Above we discussed the things to do and how to do them for the Sponsor to be successful. It is important that the Sponsor must be authentic in their communications. Further, the Sponsor needs to be comfortable with ambiguity, with not knowing everything before commencing, and with commencing with the possibility that the future may change. Having a Coach to support the Sponsor is extremely beneficial as they can assist with developing the maturing of the maturity assessment program.

1.3 How Do You Measure Success?

The traditional way of measuring success is to use Key Performance Indicators (KPIs); however, a more effective way is to have Key Behavioral Indicators (KBIs) and Key Activity Indicators. We will discuss these in detail in Chapters 3 and 5.

1.4 Key Takeaways

1. Why selecting the right sponsor is important.
2. Why setting up the nominated sponsor for success is important.
3. How to measure the success of embedding the behaviors that are delivering the results will be sustainable.

Notes

1 Marcus Buckingham, *The One Thing You Need to Know … About Great Managing, Great Leading, and Sustained Individual Success* (London: Free Press, an imprint of Simon and Schuster Ltd, 2005).
2 John Kotter, *A Sense of Urgency* (Boston: Harvard Business Review Press, 2008).
3 Oxford English Dictionary, Oxford Languages, Oxford Press, 2010.
4 Vincent Lombardi (American football coach).
5 Oxford English Dictionary.
6 Robert Collier (American motivational author, 1885–1950).

Chapter 2

Why Bother Defining Behaviors and Linking Systems and Behaviors?

Chapter Summary

Having clarity on what the desired behaviors are, deploying and then managing them is fundamental in sustaining a CI cultural transformation. Having specific measures of these behaviors in the form of KBIs drives the continual focus of refining the business systems and embedding these behaviors into the organizational DNA.

One thing that every organization has in common, whether they are commercial, not for profit, or in the public sector, is that they exist to provide results of some kind. These results may be a return to shareholders, defense of the country, saving patients' lives, or offering care and support to those in need. No matter what the results, all of them require people to deliver them. Even though we are seeing ever-increasing automation, there remains a significant dependency upon people in most organizations.

If we accept that the results we get are largely determined by the people in the organization, then the focus of any leadership team needs to be how to help their people deliver the best results. "Of course," I hear you say. "That's obvious."

The challenge becomes: How do we do this? There are numerous different approaches we can try. One view is that we document processes

16 ■ *Why Bother?*

to such a level of detail that everyone can do something in exactly the same way, every time. In the extreme, we take away the need for people to think and make choices, and instead just need them to do exactly what they have been told to do every time, all the time. If a process can really be designed this way, then it is probably best to automate it rather than expect a human being to effectively stop thinking.

At the other extreme, we can give people the total freedom to do whatever they like, however they like, so long as the results are delivered. This might sound fine and may well work in some environments, but it is likely to lead to a lot of inefficiencies and an increased risk of something going wrong.

The Shingo Institute has done much research around embedding Continuous Improvement mindsets and has developed the key Three Insights of Enterprise Excellence.

1. **To deliver ideal results you require ideal behaviors**. The results of an organization depend on the way its people behave. Whether or not an employee shows up to work in the morning will influence the results of that day. To achieve ideal results, leaders must do the hard work of creating a culture where ideal behaviors are expected and evident in every person in the organization. Figure 2.1 illustrates this insight.
2. **Purpose and systems drive behavior**. The major effect that purpose and systems have on behavior is often underestimated. Most of the systems that guide the way people work are designed to create a specific business result without regard for the behavior that the system consequently drives.
3. **Principles inform and guide people's actions and thus behaviors**. Principles are foundational rules that not only guide but direct the consequences of behaviors. The deeper people understand these values or principles, the more clearly they

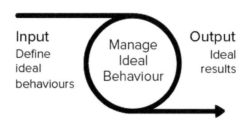

Figure 2.1 Ideal Results Require Ideal Behaviors.

understand ideal behavior. The more clearly they understand ideal behavior, the better they can design systems to drive that behavior to achieve ideal results.

2.1 Ideal Results Require Ideal Behaviors

Let us explore the first insight: ideal results require ideal behaviors.

Firstly, what do we mean by ideal results? After all, it is possible to achieve great short-term results with poor behavior. In contrast, ideal results are sustainable over the long term and are excellent in numerous different perspectives which must include social, environmental, and personal fulfilment benefits, not just financial returns.

Traditionally, we determine how we are tracking against results with Key Performance Indicators (KPI), which usually tell us what has already happened and what we need to fix after it has gone wrong. Increasingly, organizations try to identify leading indicators that help predict when a KPI may be going off track. An enormously powerful form of a leading indicator is a Key Behavioral Indicator (KBI), which aims to tell us whether the behaviors that we need to achieve the results are in place.

One way of thinking about this is that using only KPIs is like trying to drive a car while looking only in the rear-view mirror. The concept of leading indicators is not new, but they are often harder to define and measure than KPIs as they tend to have a more qualitative rather than a quantitative element. For example, how effective the coaching conversation was rather than how many were undertaken.

What we are trying to control by using KBIs are the inputs to a process, as illustrated in Figure 2.2 below:

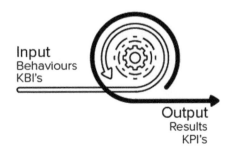

Figure 2.2 KBIs, Processes, and KPIs.

Leaders tend to focus much of their managerial effort on the outputs (what were the results today, this week, this month) and often measure success and award bonuses based on these. They may focus improvement efforts on the systems to make the results better. For example, reduce equipment downtime, implement improved flow of information or product, etc. These improvements are critical and key to driving continuous improvement in performance, but they will be limited in success and sustainability if they are not also grounded in ideal behaviors.

In summary, if leaders require ideal results then they need to define and proactively manage ideal behaviors. To do this, they need to ensure that the purpose and systems in the organization directly drive and reinforce the ideal behaviors.

This leads us to the second insight: purpose and systems drive behaviors.

2.2 Purpose and Systems Drive Behavior

2.2.1 Why Bother Defining Behaviors and Linking Systems and Behaviors?

By "system" we mean the end-to-end process rather than a specific piece of software. It is good practice to be clear on the purpose of a particular process. This may sound obvious, but all too often processes evolve in response to tactical, day-to-day issues and, before we know it, we then have an extremely complex, multi-step approach that most people find confusing.

So, what do we mean by a system? The Shingo Institute regard a system as "a collection of tools or tasks that are highly integrated to accomplish an outcome."[1] Such a system:

   has structure
   contains parts that are directly or indirectly related to each other
   have behaviors defined
   contains processes that fulfill its purpose
   has interconnectivity in terms of structure and behavior

If our purpose is to exercise total control over our people because we feel the need to micro-manage them, then this will determine the design of the process. For example, in one call center people's activities were managed every minute of the day right down to how long they spent on toilet breaks. Deviations beyond the expected standard resulted in corrective conversations with their immediate supervisor. The purpose of this system was to drive efficiency and increase output by controlling how people spent their time. However, it resulted in resentment, high staff turnover, and poor morale. People felt they were not being respected and they became totally disengaged. In this example, the system drove the wrong behaviors and led to the opposite result to that intended.

While this is an extreme example, it is quite common that the system unintentionally drives the undesired behavior. A personal example from one of the authors is given to illustrate this.

A large pub chain in the UK was offering a free pint of beer with every pie purchased. This seemed like too good a deal to miss so, as I was working away from home, I decided to take advantage of this. I ordered my vegetarian pie and pint of beer only to be told that, when my order was put into the system, vegetarian pies were out of stock. Declining the offer of a beef or chicken pie, I asked what the vegetarian options on the menu were.

There was a nice pasta dish which was two pounds more expensive than the pie, but as I was hungry I agreed to have that instead. However, much to my surprise, and the embarrassment of the server, this meant I could not have a free pint of beer because that only came with pies!

The server called over the manager and asked him to override the system so that I could have my free pint. His response was, "I'm really sorry, but I'm not allowed to do that. I want to, but the system monitors our stock levels and matches sales of pies to free pints. If we give you a free pint without selling you a pie, we will have to explain the stock discrepancy to the area manager and that's never a pleasant experience."

Realizing this was an example of a system driving behavior that people knew was wrong but they could do nothing about (and feeling hungry and thirsty), I said, "No worries," and paid for the pint and the pasta! After all, it was not their fault. No one had deliberately designed the system to deliver a poor customer and staff experience.

Systems driving undesired behavior are more common than we might think. Things just might not work as expected and as a result we get unintended consequences. When this happens, one of the most common responses is that people will find a "work around" to the system and do what they think is the right thing. How often have you heard people say something like, "Well, we are supposed to do it that way but that doesn't really work, so we do this instead?" However, they are then working against the system rather than the system supporting them to do a fantastic job. In a short space of time the system becomes discredited, there are fewer clear standards, and it is difficult to train people effectively as everyone does things slightly differently. The system then becomes inefficient and frustrating and requires a redesign. Unfortunately, unless it is redesigned with the aim of reinforcing and supporting the ideal behaviors, the cycle will start all over again. This is illustrated in Figure 2.3.

A well-designed system will drive and foster the ideal behaviors the organization requires. For example, one organization the author is familiar with has a system in place for a monthly one-on-one catch-up between a manager and each member of their team. This was measured and reviewed on a regular basis and linked to the manager's performance objectives and reviews. The intention was good, but the implementation did not achieve the desired results. Instead of fostering a culture of support and personal development, the measure drove a behavior of "ticking the box" to complete the target number of reviews—often in a rapid rush of quick-fire conversations in the last couple of days of each month.

An assessment highlighted that, while the KPI was being met, the system was not delivering anything of real value and had largely become discredited. Nevertheless, the organization realized the high value of getting the system right, so it was redesigned. They decided that the customer of the process was the employee and that the customer should evaluate the effectiveness of this system. Instead of giving the managers a target of how many one-on-ones to complete, the measure was changed to a KBI—what percentage of a manager's team rated their one-on-one as highly valuable in a simple online and anonymous Net Promotor Survey.

The system was further refined to provide support to managers on how to have better conversations and how to celebrate successes. Group peer coaching sessions, sharing lessons learned, and how to improve the quality of the conversation were also established as standard work. Employee engagement in the process increased dramatically, with people looking forward to their one-on-ones and holding their managers accountable for ensuring a timely and valuable two-way conversation.

One of the most common reasons that systems do not deliver the intended outcome is a failure to first clearly define the purpose of that system. In the example above, the purpose became to do one-on-ones rather than to use the one-on-one conversation to develop both the manager and the employee while also increasing overall engagement levels.

We recommend that every system should have a clearly defined purpose, otherwise, how can we judge whether it is effective or not? Hines and Butterworth in *The Essence of Excellence*[2] detail four core systems necessary

Why Bother Defining Behaviors and Linking Systems and Behaviors? ■ 21

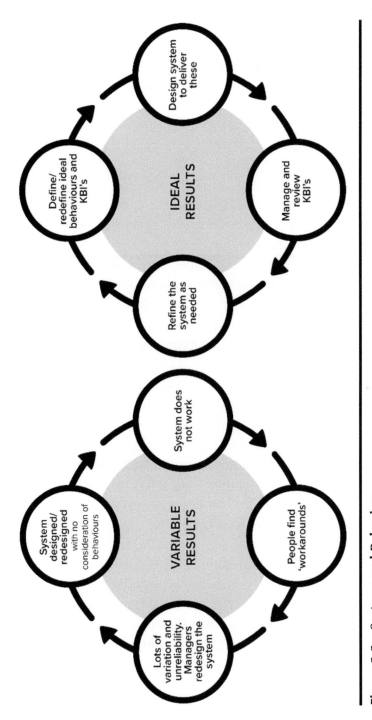

Figure 2.3 Systems and Behaviors.

for organizations to sustain a culture of continuous improvement. These systems are:

- Strategy and Behavioral Deployment
- Continuous Improvement
- Leader Standard Work
- Learning and Development

An example of a purpose statement for one of these systems (Learning and Development) is given below.

The What: Learning and Development is a system that develops a learning culture and continuously develops its people aligned to the needs of the organization.

The How: By identifying Learning and Development opportunities at all levels; focusing on the improvement of work competencies, improvement competencies and culture change competencies; undertaking pull-based coaching to improve competence, performance, and culture; ensuring that Learning and Development competencies are embedded and lead to changes in behavior and tangible results; and recognizing people for their learning achievements.

The Why: So that the whole organization maximizes its human potential, with the members taking initiative.

Having defined what the system is, how it will operate, and why it exists, the authors suggest it is then useful to create a high-level systems map showing the inputs, the key elements, and the outputs. An example of the Learning and Development system is shown in Figure 2.4.[3]

One of the other systems that Hines and Butterworth[4] explain in detail is the Strategy and Behavioral Deployment system. One key aspect of this system is to determine the ideal behaviors required to deliver on the organization's higher-level purpose and then to deploy these behaviors. Figure 2.5 summarizes the behavioral deployment process.[5]

So far so good, but we now need to integrate these ideal behaviors into our systems. Deming teaches us that "a bad system will defeat a good person every time."[6] In other words, no matter how well we define the behaviors the system will either support them or destroy them.

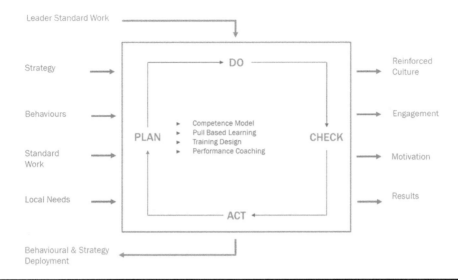

Figure 2.4 The Learning and Development System.

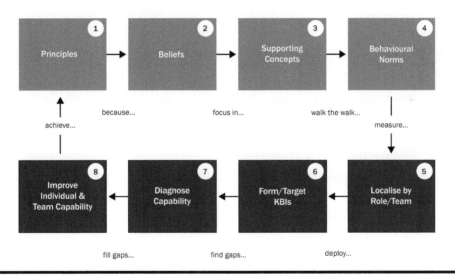

Figure 2.5 Behavioral Formation and Deployment.

The starting point we recommend is to define the business-level ideal behaviors. How to do this is described in detail in *The Essence of Excellence*[7] and some examples of these behaviors are given below.

"We will always show respect to each other by taking time to really listen, understand, and show appreciation for each other's inputs."

"We will always strive to ensure that our actions achieve the best possible outcome for our customers and each other."

Once these have been established, one approach that can be used to integrate systems and behaviors is illustrated below.

Step 1: Identify the key business systems.
Step 2: Agree on the prioritization of which system(s) need to be addressed first. For example, which ones will have the biggest impact on strategic goals or which ones are known to be not working well.
Step 3: Agree on a purpose statement for the priority systems.
Step 4: Agree on a high-level systems map.
Step 5: Agree on the results based/output Key Performance Indicators (KPIs) and targets for each system.
Step 6: If useful, at this stage undertake a Value Stream Mapping (VSM) activity on the system (it may not be useful, for example, if a VSM activity has recently been completed or the system has only recently been redesigned).
Step 7: Use the behavioral deployment process to get the teams that use the system to define how they will behave at a local level to ensure KPIs are achieved.
Step 8: Define the Key Behavioral Indicators (KBIs) that will measure the agreed behaviors.
Step 9: Ensure there is a structured feedback loop that responds to what the KBIs are telling us; for example, if the system is not supporting a particular behavior or even driving the wrong behavior.
Step 10: Ensure there is a corrective action process to constantly review and refine the system driven by the feedback.

Some good questions to ask when undertaking this work are:

1. How do we design this system, or what changes do we need to make to the system to drive and nurture the ideal behaviors?
2. How can we design the system so that it is harder to behave in the undesired way than it is to behave in the ideal way?
3. What are the KBIs we use to understand if this system is driving and nurturing the ideal behaviors?
4. How do we help teams at a local level develop their KBIs? For example, the finance team is likely to have different requirements to the operations team.
5. What is the feedback loop which will enable us to refine and develop the system and the KBIs?

In summary, systems and behaviors are two sides of the same coin. If we have a good system but poor behavior, then the system decays over time. If we have a poor system but good behaviors, this leads to a lot of hard work, rework, frustration, excess working hours, and waste. If we have a good system and ideal behavior, then this leads to a mutually reinforcing feedback loop where system and behavior support each other and enable further improvement.

As we have demonstrated, implementing a system in isolation of behaviors can be very limiting for both the effectiveness of the system and the morale of the people in the organization. Therefore, defining ideal behaviors is critical. This leads us to the third insight: principles inform ideal behavior.

2.3 Principles Inform Ideal Behavior

A behavior is something you can see and/or hear. An effective way to think about this is that a behavior must be something that you could video—demonstrating what future-state behaviors (ideal) would look and sound like.

While it may sound simple to define behaviors, it is very helpful to have something to use as the basis from which to work. Many organizations have a set of company values; these can be a good place to start. Defining behaviors for those values is an effective way to start to bring them to life. It is important for leaders to be especially clear when they describe these behaviors that they are something that can be observed (seen and/or heard).

To help develop ideal behaviors, it is essential to use values or principles to inform our thinking. One of the leading examples of this approach comes from The Shingo Institute at Utah State University. They have developed a behavioral-based approach to improvement that became known as Enterprise Excellence. This approach is based on the five elements of the Shingo Model, namely:

- a set of ten "Guiding Principles" deployed into
- a set of "Systems" (such as Strategy Deployment & Order Fulfilment) overlaid by a behavioral-based "Culture" that enables the Systems to work effectively and sustain themselves supported by
- a set of "Tools" derived from the defined Systems (not the other way around) which lead to
- a set of appropriate and sustainable "Results"

26 ■ *Why Bother?*

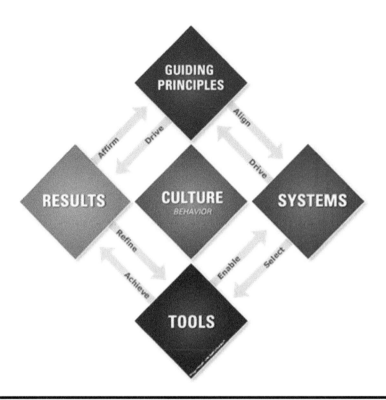

Figure 2.6 The Shingo Model.

The Shingo Institute[8] suggests that the secret to create a sustainable Enterprise Excellence is in developing a principle-led approach and deploying this into the organization. In this sense, they suggest that a principle is a foundational rule that has an inevitable consequence. It is universal and timeless, self-evident, and governs the consequences of following it or not.

The Shingo Model is reproduced in Figure 2.6. The organizational culture demonstrated and managed through the behaviors is shown at the heart of the model. These behaviors are informed by the Guiding Shingo Principles.

2.4 The Shingo Principles

The Shingo Principles are categorized into three dimensions:

Enterprise Alignment
Continuous Improvement
Cultural Enablers

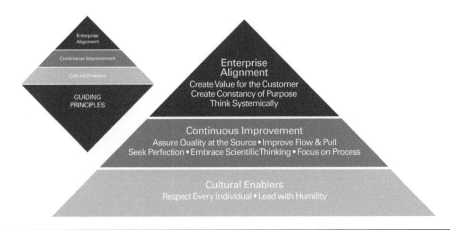

Figure 2.7 The Shingo Dimensions and Principles.

Each of these dimensions has several principles and these are shown in Figure 2.7.

Below, by way of example, we give a high-level summary of the principles for Enterprise Alignment with some example behaviors.[9]

2.4.1 Enterprise Alignment Principles

2.4.1.1 Results — Create Value for the Customer

The best and only assumption an organization should make about customer value is to assume that they do not really know what the customer values. Time and again we see organizations making changes to products and processes without first understanding what the customer values. Without this understanding, we risk changing the wrong things and either removing something the customer values or adding something they do not really care about. But how do we know if it will make the experience better for our customer?

A deep understanding of customer value is critical to drive business excellence and innovation. We believe that traditional survey approaches fail to identify customer value. According to Womack and Jones,[10] the first lean principle is to understand what customers value. At first, this may seem an obvious starting point for any organization wishing to be successful in whatever products or services they are seeking to supply. However, it is often overlooked, with organizations assuming they know what their customers value.

Table 2.1 Create Value for the Customer Example Behaviors

Create Value for the Customer	Example Behavior
Leader	Ensures that what the customer values is embedded in all decision making.
Manager	Constantly ensures their team understands how their actions relate to what the customer values.
Associate	Uses what the customer values to inform decisions and priorities.

At a high level, changing customer value is the underlying driver of why we need to develop a culture of continuous improvement in every organization. If we do not have a way to understand customer value and we continuously implement changes to meet the changing value criteria, we run the risk of becoming irrelevant to our customers.

To quote the Shingo Institute:

> Ultimately, value must be defined through the lens of what a customer wants and is willing to pay for. Organizations that fail to deliver both effectively and efficiently on this most fundamental outcome cannot be sustained over the long term.[11]

Table 2.1 gives some examples of behaviors associated with the principle.

2.4.1.2 Constancy of Purpose

The principle of Constancy of Purpose is about aligning everyone in the organization with a common purpose. The simplest analogy to illustrate this principle is the rowing boat in a race—it goes faster and has a much happier crew if they are all rowing towards the same destination with the same rhythm and common goal to get to the finish line as fast as possible and win the race. Chaos ensues if someone decides their rhythm is better or more important than the rest of the team or just decides to do their own thing.

The purpose of the organization needs to be constant over time—constant as it is deployed down through the organization and constant across the functions and departments. In other words, there is a consistent understanding at all levels and in all areas of why the organization exists, what it is trying to achieve, and what are the strategic goals.

Table 2.2 Constancy of Purpose Example Behaviors

Constancy of Purpose	Example Behavior
Leader	Frequently communicates and reinforces a clear and common purpose.
Manager	Demonstrates actions that ensure that the purpose is understood and deployed to all team members.
Associate	Always discusses the purpose at the start of team meetings.

Without this constancy of purpose, any organization becomes a very confusing place to work; priorities change frequently and people reactively lurch from one "crisis" to the next, quickly losing sight of what the organization is trying to achieve.

Table 2.2 gives some examples of behaviors associated with the principle.

2.4.1.3 Think Systemically

While Constancy of Purpose focuses on the vertical alignment of the organization, the principle of Think Systemically focuses more on the horizontal alignment. Customers do not care about departments and value rarely flows through just one department in any organization. However, silos are often so strong they prevent the efficient flow of value and it is easy for people to lose sight of this.

There are several ways that organizations seek to apply the principle of Thinking Systemically. One of the options is to change the emphasis of KPIs and targets so that they do not reward departmental performance but instead reward business performance and collaboration. Another approach is to appoint high-level process owners whose role is to oversee the end-to-end process and make the call on any perceived department conflicts or barriers that obstruct value flowing smoothly.

To quote the Shingo Institute:

> Through understanding the relationships and interconnectedness within a system we are able to make better decisions and improvements. … As we see how and why everything is connected to, or part of, something it helps us to better understand, predict and control outcomes.[12]

Table 2.3 Think Systemically Example Behaviors

Think Systemically	Example Behavior
Leader	Proactively encourages collaboration across departmental boundaries.
Manager	Always focuses on optimizing the end-to-end process across the business not just improving their own department at the detriment of other areas.
Associate	Understands how their role contributes to the end-to-end process and seeks to optimize the whole.

Table 2.3 gives some examples of behaviors associated with the principle.

In summary, there is a high-level, four-step approach to defining and implementing ideal behaviors and ensuring that these are integrated with our systems to deliver ideal results. This is:

1. What results do we want?
2. What behaviors are needed to achieve these results?
3. How does the system need to be designed to drive and support these behaviors?
4. What KBIs will we use to measure behaviors and check if the system is working effectively?

Step 4 will be discussed in more detail in Chapter 5.

In this chapter, we detailed the importance of establishing ideal behaviors and integrating these into our systems. We have provided some examples and approaches on how to do this, as we believe it is essential that for assessment systems to be effective they must focus on behaviors. In other words, if we really want to understand the maturity of our CI culture we need to have an assessment system that measures behaviors. How to design this type of assessment system will be explained in Chapter 4.

2.5 Key Takeaways

1. You cannot manage behaviors unless you clearly define and deploy them.

2. A wide range of techniques are needed to constantly reinforce and encourage ideal behaviors.
3. Systems need to be continuously assessed and updated to ensure they are supporting and driving ideal behaviors.

Notes

1. www.shingo.org.
2. Peter Hines and Chris Butterworth, *The Essence of Excellence: Creating a Culture of Continuous Improvement* (Caerphilly: SA Partners, 2019).
3. Hines and Butterworth, *Essence of Excellence*, 167.
4. Hines and Butterworth, *Essence of Excellence*, 36–84.
5. Hines and Butterworth, *Essence of Excellence*, 46.
6. "A Bad System Will Beat a Good Person Every Time"—The W. Edwards Deming Institute. www.deming.org.
7. Hines and Butterworth, *Essence of Excellence*, chapter 3, 36–84.
8. www.shingo.org.
9. For a much deeper understanding of the Shingo Principles, we highly recommend the *Shingo Model Handbook* which can be downloaded free from www.Shingo.org and the various highly interactive Shingo Institute workshops.
10. James P. Womack and Daniel T. Jones, *Lean Thinking: Banish Waste and Create Wealth in Your Corporation* (New York: Simon & Schuster, 1996), 16.
11. *Shingo Model Handbook* (Shingo Institute Jon M. Huntsman School of Business Utah State University, 2017), p. 36.
12. Ken Snyder, "A Look at 'Think Systemically'," https://shingo.org/a-look-at-think-systemically/.

Chapter 3

Why Bother Assessing and Managing Behaviors?

Chapter Summary

Continuously maturing behaviors for existing employees and onboarding new employees into the desired behaviors is essential to sustaining a CI culture. There is a wide range of activities needed to manage and embed ideal behaviors. A behaviorally based maturity assessment is key to enabling us to track progress and identify actions needed to correct slippage or progress to the next level. Assessments that only review tools or systems will not support the development nor sustain the CI culture of the organization.

3.1 Assessments

Various forms of assessment have been used for many years in the CI world. Just as our understanding of what makes success more likely has evolved from tools to systems to behaviors, so assessments need to evolve as well.

What we have come to realize is that, instead of starting with tools, we really need to start with the desired culture and behaviors, then design systems that reinforce these, and then select the appropriate tools that support our people.

Step	No	Description	Not Found	In some areas	Common-place	Very Typical	Every where	Comments/Observations
Sort (Organisation)	1	Workbenches and cupboards are clear						
	2	The area is clear of any existing or potential safety hazards						
Set in Order (Orderliness)	3	Visual Aids (e.g shadow boards) are used throughout the area						
	4	There is a place for everything and everything is in its place						
Shine (Cleanliness)	5	All floor areas are to the defined standard condition						
	6	All machines /equipment are in the optimum condition						
Standardise (Adherence)	7	All the necessary Standard Operating Procedures are visible						
	8	All items are correctly labeled and identified as required						
Sustain (Discipline & Improvement)	9	There are regular 5S Audits undertaken						
	10	Standards are constantly improved						
		TOTAL SCORE						

Figure 3.1   A 5S Audit Extract Example.

Historically, many assessments focus on the application of a particular tool. For example, the tool 5S, such as the example of an extract from a 5S audit shown in Figure 3.1.

These may well be useful in the preliminary stages of implementation or to drive standards at a local level, but they tend to focus on the *what* rather than the *how* and the *why*. As such, they have limited impact when driving a sustainable cultural change. There is also a tendency for them to be viewed as audits that are conducted to catch people out and thus risk driving undesired behaviors. It is not unusual for teams subjected to these kinds of audits to undertake an activity for the purpose of the audit rather than for any perceived benefit, leading to a lack of engagement and poor sustainability.

A wide range of assessments has been developed at the systems level, and one widely used example is the process maturity assessment. There are several variations on the approach, but all essentially assess processes at various levels of maturity and allocate a score. They are based on the process maturity scale, with Level 1 typically being no real process in place through to Level 5 being a process that is well established and being fine-tuned or optimized.

This is an extremely useful approach and highlights areas where there are gaps that need to be addressed. It can be particularly useful in organizations where compliance to external standards is critical, such as banks. However, rarely does it explicitly focus on what behaviors are required, nor does it usually assess how effective a particular system is at driving ideal behaviors.

3.2 Purpose of Behavioral Assessment

One of our key learnings over the years has been that the sustainability of a CI culture depends not on the tools being used but on the behaviors established and managed through constant reinforcement.

While establishing the desired behavioral norms or standards is a critical starting point, it is equivalent to sowing seeds in your garden. You can put them out there and leave them to it—or, as some would say, "launch and leave." If you are lucky some seeds will grow and flourish, but others will start to sprout then wither away and some will never even germinate. The chances of success will be hit and miss. Embedding behaviors is exactly the same—we need to constantly manage and nurture desired behaviors if we want them to flourish.

If leaders want a CI culture to grow and thrive, then they need to proactively manage and nurture the agreed behaviors; regular reinforcement and support for what is going well and dealing quickly with the weeds that sprout up are critical. If leaders "launch and leave," then the culture will quickly revert to local level interpretation and the desired business-wide culture will die away. To quote Professor Edgar Schein:

> The only thing of real importance that **leaders do is to create and manage culture. If you do not manage culture, it manages you**, and you may not even be aware of the extent to which this is happening.[1] (Emphasis added)

The reinforcement of the ideal behaviors can take many forms, ranging from a simple one-to-one recognizing the desired behaviors, to business-wide celebrations. Some of the most effective forms of reinforcement are detailed below.

1. Nudges.
2. Forming habits.
3. Recognition in various forms.
4. Look, Listen, and Learn (Gemba) walks.
5. Systems constantly reviewed and improved to ensure they drive and support the desired behavior—reference Shingo.
6. Quickly dealing with behaviors that are not desired.
7. Role modelling the ideal behaviors.

8. Respect.
9. The recruitment process (case study from a leading Australian financial services business).
10. Strategic level behavioral assessment system—this will be detailed in Chapter 4.
11. Managing Key Behavioral Indicators—case study from Peter Hines et al. in Chapter 5.
12. The types of conversations people have—paper from Kevin Eyre in Chapter 6.

Our view is that a combination of all the above is needed and that a well-designed behavioral assessment system can be used to support each form of reinforcement. Each of these forms of reinforcement is expanded upon below.

3.2.1 Nudges

Nudge is a concept in behavioral economics, political theory, and behavioral sciences that proposes positive reinforcement and indirect suggestions as ways to influence the behavior and decision-making of groups or individuals. The nudge concept was popularized in the 2009 book *Nudge: Improving Decisions About Health, Wealth, and Happiness*.[2]

A nudge makes it more likely that an individual will make a particular choice, or behave in a particular way, by altering the environment so that automatic cognitive processes are triggered to favor the desired outcome. Nudges are minor changes in the environment that are easy and inexpensive to implement.

An example of nudge being applied to influence behaviors in the workplace is explored in *4+1: Embedding a Culture of Continuous Improvement in Financial Services*.[3] The case study company introduced four questions across the whole organization. These were used regularly in team briefings and decision-making and in performance conversations. Teams were also encouraged to openly discuss the questions, and leaders and managers regularly asked questions.

The questions were:

1. **Am I focused on activities that add value for the customer and the business?**

This is all about provoking people's thinking to question why they are doing some of the crazy things that frustrate them—often the things they know they should not be doing.

2. **Are our processes efficient and performed consistently across the team?**
This is about standardizing the things people should be doing, removing and reducing waste, and making processes consistent every time they are performed.

3. **Are we focused on the right activities at the right time with the right skills?**
This is about understanding customer requirements and the requisite people skills—making sure we get the right people working on the right thing with the right skills at the right time.

4. **Am I personally contributing to continuous improvement every day?**
This question brings the other three questions together, with the aim to create an environment where everyone continues to learn and apply the improvement habits. If people use these questions on a regular basis, they change the way they look at their work. If the answer to any question is "no," then there is an opportunity for improvement.

These questions are nudging people to think differently and, if used on a regular, long-term basis, will drive a new set of behaviors and support the development of new habits.

3.2.2 Forming Habits

A true understanding of how people's brains work and, at the same time, understanding their behaviors is crucial for creating an alignment to a standard set of behaviors. John Medina and David Rock[4] (*Coaching with the brain in mind* and *Quiet leadership*) examined the development of habits that drive the desired behaviors that collectively create a culture of continuous improvement.

1. **The brain is a connecting machine**—It is phenomenal at making connections, especially around trends in data and targets as well as abstract connections of cause and effect.

2. **No two human brains are alike**—Each of our brains has developed uniquely due to various experiences, external influences, and family/societal influences.
3. **Constrained conscious processing capability**—The brain filters the amount of information it requires for processing by using previously formed frames of reference to categorize and remove less important or irrelevant information. Most of the brain's power is in the subconscious part of the brain (where habits are located), whereas conscious thoughts occur in the newer part of the brain (the prefrontal cortex) which, by comparison, is volume constrained on the brain's processing power.
4. **The brain does not unlearn**—The brain does not unlearn things, except due to medical trauma. The brain creates new neural pathways that drive new thinking and this informs new behaviors. Initially, when we are learning new things, we must be conscious about the new thing we want to do. This information stays and is processed in the prefrontal cortex or conscious part of the mind. Then, over time, this neural pathway becomes stronger as it is performed repeatedly and eventually moves into the subconscious. A true habit is something we do without having to consciously think about it.
5. **Stressed brains do not learn or listen**—Under stress or perceived danger, blood (containing much of the brain's fuel) flows away from the prefrontal cortex to the major muscle groups in preparation for fight or flight, reducing the ability of the individual to think rationally, make connections, and, most importantly, listen. In the threat state, the brain's ability to solve problems and collaborate effectively is severely reduced.

A recent discovery that the human brain can change its own structure and function through thought and activity is perhaps the most important alteration in our view of the brain in four hundred years. Known as neuroplasticity, it has the power to allow us to change, but it can also lock us into behaviors too. Neuroplasticity has enabled the blind to see, the deaf to hear, the learning disabled to learn, and an 80+-year-old to sharpen their memories to that of a 55-year-old. Norman Doidge's book *The brain that changes itself*[5] covers neuroplasticity in detail.

The brain automates everything it can to conserve energy. Change requires vastly more energy to pay conscious attention! An example is learning to ride a bike—we wobble around, fall off, and look stupid.

It takes attention, practice, and correction, and then more practice and correction to move from beginner to competent. Add more practice—up to around 1000 repetitions—and the behavior becomes automatic due to neuroplasticity, as explained in four stages:

1. **Unconsciously incompetent (unskilled)**—You do not know what you do not know. You are unaware of your ability or lack of in relation to a skill.
2. **Consciously incompetent (unskilled)**—You now know what you do not know and begin taking the initial steps towards becoming competent.
3. **Consciously competen**t—You can demonstrate the skill, but it still requires your full attention and is very effortful. You will be using your prefrontal cortex, or PFC (just behind your eyes, which is the most recent "add-on" to our evolving brain), and this part of the brain is very energy demanding. The brain consumes around twenty percent of our body's available energy and yet it weighs only just over three pounds.
4. **Unconsciously competent**—This fourth stage is like riding a bike without having to think about it, and possibly chatting with a friend at the same time. There are different levels within each stage and at the unconsciously competent stage, you might be proficient, an expert, or a master for example. Now, instead of engaging the PFC (the energy guzzler), the basal ganglia, which is far more energy-efficient, takes over.

Is it any wonder that many people resist change in the preliminary stages? However, dealing with change is much easier if you can predict what is ahead (the brain desires certainty, so being able to predict what lies ahead allows it to feel safe). Safety is the brain's number one priority. It is much easier to predict what is ahead when you have a model in your head and you can assess where you are and where you are heading. Neuroplasticity is in action whether we like it or not, either reinforcing/ locking in behaviors that become unconscious habits—which either serve us or do not—or through self-directed neuroplasticity, allowing us to choose and make the changes we want to make, initially through massive practice and correction.

What single change about your thinking or behavior will you now work on to take advantage of self-directed neuroplasticity?

A child learning to play piano scales for the first time is a good example. They tend to use their whole upper body—wrist, arm, shoulder—to play each note. Even the facial muscles tighten into a grimace, but with practice, the budding pianist stops using irrelevant muscles and soon uses only the correct finger to play the note. They soon develop a lighter touch and, if they become skillful, they develop grace and can relax when they play. The demand drops from a massive number of neurons to an appropriate few, well-matched for the task.

This more efficient use of neurons occurs whenever we become proficient at a skill and explains why we do not quickly run out of (neural) map space as we practice or add skills to our repertoire. Trained neurons fire more quickly, process faster, and recover quicker, ready to fire again. Faster neurons = faster thoughts, a crucial component of intelligence and vital for the pace we are challenged to operate at today.

Plenty of practice of the new behavior is required, with appropriate increases in complexity, to move from self-conscious novice (consciously incompetent) to expert (unconsciously competent).

So, in relation to the new behaviors and habits you want to develop and utilize, are you prepared to do what it takes to go from where you are to where you want to be? Will you put in place support systems that encourage you to meet your commitment to practice, be open to mistakes and correction, and apply any new learning?

To form habits and create new neural pathways, there are two primary requirements—attention and positive feedback to that new connection. Attention required is both quality and quantity of attention—called attention density. Ideally, if you would like people to do things in a new way you need to pay regular attention to the small changes, build on those small changes, and then create a new habit. Often managers do not praise a beginning or newly formed skill as the behavior is only partway to a high-performance habit, but that is precisely when we need to pay attention to and praise that new behavior.

The positive feedback required can be in the form of comments, such as, "Well done for trying X," or, "That is a great start towards X," or, "Now that you have started that new habit, which is great, how can you practice it on a more frequent basis?"

The combination of attention and positive feedback could be summarized as "Mind the GAP"—be "Mindful" of the habit and what is occurring for the person (i.e., the resistance they are overcoming or the fear of getting it wrong); then focus on the "Goal" (what outcome you are after and help remind the person of this outcome); then pay "Attention" regularly—quality and quantity, daily or more, depending on the task; and provide "Positive" feedback, often in various forms, to the new behavior, even for small changes or increments or just for trying.[6]

In *4+1* Jones, Butterworth, and Harder[7] go into detail about the neuroscience underpinning several CI habits. One example linked to ideal behaviors at team "huddles," or visual management board meetings, is given below.

> **Huddles**—Human beings are social creatures by nature. Creating safe spaces for people to celebrate wins, acknowledge one another, ask clarifying questions, and share and constructively evaluate ideas in a group setting can support and optimize team members' brains. Feeling safe and supported (even when challenged) allows the brain to direct blood flow to the prefrontal cortex, allowing the individual to better attend, focus, listen, interact, and create the conditions to facilitate insight.

One way of thinking about how to create constructive social interactions is to use the SCARF model by David Rock,[8] which can help to make those interactions and, more specifically, huddles more successful. Science is now discovering that social pain—the pain of feeling rejected, for example—shows up in the same area of the brain as physical pain.

Unlike many other motivational theories, the SCARF model is quite modern. SCARF is an acronym:

Status—the relative importance of self to others
Certainty—being able to predict the future
Autonomy—a sense of control over events
Relatedness—a sense of safety and belonging with others
Fairness—a perception of fair exchanges between people.

Behind this is the idea that a person's brain will activate behaviors to minimize perceived threats and maximize rewards, and that the brain reacts in the same way to social needs as to our primary needs of food and water. If a stimulus—for example, a huddle—is associated with positive interaction, positive emotions, and rewards, a person will probably move towards it physically or emotionally. But if a stimulus is associated with negative interaction, negative emotions, or punishments, it will be perceived as a threat, and the person will probably move away from and avoid it. That all sounds straightforward.

The SCARF model can be used to plan the setup and running of huddles in such a way that threats are minimized and rewards are maximized in each of the five areas, or domains, of Status, Certainty, Autonomy, Relatedness, and Fairness.

If a person feels that they are being threatened, their emotional brain—particularly the amygdala—will work quickly to protect them, instigating the flight-fight response. This will reduce their capacity for rational thought, to make good decisions, to solve problems, and to collaborate—which are the things we want teams to do.

What makes the SCARF model so useful is that it identifies the five domains that activate the primary reward or primary threat circuitry in a person's brain.

If we want to embed ideal behaviors, we need to make them a habit so that the undesired thing becomes much harder to do than the ideal thing.

3.2.3 Recognition in Various Forms

Recognition is a particularly useful form of nudging and helps to create a safe learning environment that is essential to form new habits. In *The Essence of Excellence*, Hines and Butterworth[9] identify recognition as a key element of a leader-standard work system.

There has been a great deal of research into recognition. Some of the most practical findings have come from interviews conducted by the O. C. Tanner Institute,[10] which conclude that recognition:

> **causes great work**—the highest cause of great work was recognition (seven percent of people interviewed);
> **increases engagement**—seventy-eight percent of people who are strongly recognized are highly engaged against thirty-four percent who are not strongly recognized;
> **encourages innovation and productivity**—forty-one percent of employees would choose recognition for going "above and beyond" as their number 1 choice in encouraging them to innovate and be productive, compared to only thirty-two percent who would choose a five percent pay rise;

improves trust and managerial relationships—fifty percent of employees said that recognition for ongoing effort would be the most important thing in improving their relationship with their manager, compared to only twenty-two percent who cited a five percent pay rise; and

attracts and retains talent—the average time people stay in companies with service awards is 6.7 years compared to 4.7 years in those that do not.

Recognition is about creating good feelings and sustainability; it is about celebrating success and it is, above all, about reinforcing ideal behaviors. If things do not go well, they should be seen as a learning and development opportunity and managed accordingly.

The importance of recognition is reinforced by research conducted by the Shingo Institute regarding employee engagement. The work found that five key factors had the biggest impact on employee engagement. These are summarized in the article *Best Ways for Manufacturers to Boost Employee Engagement*[11] on the Shingo website as follows:

1. Employee perceptions of the opportunities available for personal development in the form of cross-training and other learning that might lead to job variety, value to the company, and possible advancement.
2. Consistent publicly expressed appreciation from leaders and managers for ideas, work, and other contributions from employees.
3. Employee perceptions of the relative availability of the tools, training, direction, and knowledge required to meet the work demands placed upon them.
4. The extent to which an employee is able to make meaningful decisions regarding their work.
5. Understanding of how their tasks help the organization accomplish its goals.

All these elements should be considered for inclusion in any CI assessment.

Recognition works and the simplest, easiest, and most repetitive methods are generally the most effective. There are many ways of recognizing our people, ranging from a simple thank you to publicly celebrating the results of a CI assessment.

3.2.4 Look, Listen, and Learn (Gemba) Walks

Most commonly referred to as Gemba walks, this leadership habit is crucial to embedding desired behaviors. While it is important to understand what Gemba walks are and how to undertake an effective one, the most critical place to start is to answer why to do them at all. In other words, let's be clear on the purpose before we jump into the activity.

The purpose is not the walk itself but rather what we see and hear on the walk and what we learn from our observations. But even more importantly—what actions are generated as a result of what we learn?

Before starting any walk, always pause and ask, "What is the purpose of this walk?"

There may be many reasons for doing a Gemba walk: to check the process; to review improvement activity; to see if task-level, standard work is being followed; to see if people are communicating in an effective way; to gauge the state of the workplace organization; or to see how behaviors are managed. Listed in Table 3.1 is a summary of the types of things that you should be able to see on the walk, as well as what people should know.[12]

3.2.4.1 What?

Gemba walks are checking processes that lead to a greater understanding of what's really happening in the workplace and are, therefore, the foundation for further learning, development, and improvement. They help distinguish between the process and the people. They provide a way to help understand why the existing processes may not be capable enough and help people to understand that they are not to blame. It is about asking yourself the "5 Whys," not the "**5** Whos."

What Gemba walks provide is the opportunity to learn firsthand what is really happening in the workplace; not so much whether targets are being met, but what support people need to help achieve those targets. We need to look at what is happening and listen to what people are saying, both to us and to each other. Typical things we want to learn could include, for example:

- Do people understand the business priorities?
- Do people feel they get the support they need to achieve their targets?
- Are people skilled in how to solve problems in a structured way that addresses the root cause?

Table 3.1 Gemba Walk Themes

Walk Focus	What you should see	What people should know
Communication	Daily shift meeting agenda visible on the team info centre	How often does your team meet as a group
	Where applicable, info from other shifts is displayed	Is it a regularly scheduled meetings or just once in a while?
	Meetings happen at all management levels	How do you know what topics you'll cover in any given day's start-up meeting?
		Do you lead or attend any daily meetings?
		What are they?
Behaviours	A current list of behavioural norms of the business	What are the behavioural norms used in this business?
	A current list of specific behaviours for team members, team managers and manager	What are the specific behaviours that you developed in this team?
	A method of recognising these behaviours	how are they assessed?
	A method of improving behaviours where there are gaps	What happens if they are applied?
		What happens if they are not?

- Are people living and breathing the organization's ideal behaviors?
- Are our systems making it easy to do the job in the right way?

The high-level process can be described as a six-step approach (Bremer, 2016), as illustrated in Figure 3.2 below.[13]

If this approach is taken, then the walker will understand the current target and current state of the topic under review. They then discuss the desired improvement with the people they meet, any specific obstacles, what has been learned, and what the next steps are. This allows for appropriate coaching conversations about how to reach the target state and whether this target should be extended.

46 ■ *Why Bother?*

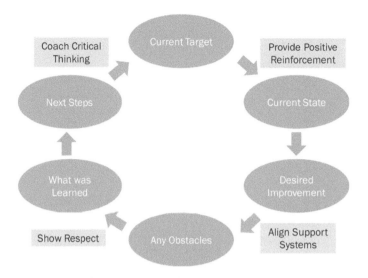

Figure 3.2 Six-Step Approach to Gemba Walks.

We often hear, "We do Gemba walks—I visit the shop floor every day." While this is a good first step, it does not mean that there is an effective Look, Listen, and Learn process in place. Unfortunately, the activity can be driven by the wrong metric, such as a target of how many walks each manager must do but without any qualitative check on their effectiveness. In these scenarios, people are essentially using a tool but risk undermining its effectiveness and credibility. Consider the example below from *The Essence of Excellence*:[14]

A senior manager has decided to do a weekly Gemba walk to the shop floor. He visits a local team leader and notices that the team board has a lot of 'red ink.' Many of the KPIs are below target and most of the projects are behind. Without delay, he starts angrily shouting at the team leader about what a poor job he is doing and that when he comes back next week he expects to see no 'red ink,' with all targets being met and projects up to date. The whole exchange takes no more than five minutes. The likely short-term effect is a highly demotivated individual who may well pass this sentiment on to his team. The longer-term effect is that problems will be hidden, so the chances of improvement are low.

Having learned from his mistake, the senior manager visits another team the week after and finds a similar amount of 'red ink' on the board. Instead of shouting, he calls the second team leader over and conducts a ten-minute monologue about what the biggest problems are, what needs to be done, and exactly how to go about it in the period before he visits again in a week's time. Again, if we consider the consequences, in the short term there may well be some work undertaken by the team leader and team, but it is likely to be half-hearted, poorly executed, and unlikely to be sustained, as it was not their idea, may not address their issues and was conveyed in an authoritative way. The longer-term effect is equally poor as the team leader has been taught not to think for himself and even if he does, he won't be allowed to implement his ideas. The result is a slow improvement, with generally suboptimal results and poor sustainability.

A week later, the senior manager has the third attempt, this time taking a very different approach. He visits another team leader with the same 'red ink' scenario. He starts by asking how things are going and what the major two or three issues seem to be. The team leader responds by talking through three areas. The senior manager takes one of these—the area that due to his experience he considers the most pressing—and asks what the team leader thinks might be done about it. The team leader responds with a few ideas. The senior manager picks up on the first one and they proceed to discuss how this might be done. At the end of a 30-minute discussion, the team leader has a clear plan for his top two projects, what to do and how to do it. The senior manager signs off with a question about whether any more support is needed. The short-term result is that the team leader goes away all fired up, conveys this enthusiasm to the team and rapid progress is made. The longer-term result is that the team leader has been coached and will have a much better idea of what to do next time and may also start to adopt a similar approach with their own team.

Three quite different results, all from a Gemba walk. However, the three approaches need an increasingly long time for the conversation, at least for the first time around. Although many of us know that the latter will produce the best result, we do not do it that way as we do not have time. The moral is if we do not make the time we will never have time because we will be firefighting the problems caused by the first two scenarios.

The style of leader-standard work is as important as the content. Mann notes that "Gemba walking, traditionally focused on the technical side of Lean (is inventory being pushed or pulled; is standard work balanced to takt time?), is indispensable in learning Lean management and especially in maintaining it." However, we believe that it is as, or more, important to check the behaviors of teams and their leaders, because it is their projects and results. Of course, this is much easier to do if you have a good "behavioral and strategy deployment" system in operation. This is described in detail in chapter 3 of *The Essence of Excellence* book.

Look, Listen, and Learn is a useful way to remind yourself of how to undertake a Gemba walk. It is something that takes a lot of practice and it is very difficult to get it right on the first attempt.

Many leaders struggle to make the time to undertake Look, Listen, and Learn walks, and sometimes it is necessary to create the time by changing working practices. An example of this is shared below.

Visiting a company I hadn't been to for several months, I checked in for my appointment with the CEO. I was taken to his office close to the entrance of the building and noticed a small queue of people outside waiting to see him. Eventually, it was my turn and we got talking about how things were going. It soon became apparent that he had become so wrapped up in working in the business (largely from his office) that he had not been on the shop floor for over three months. The queue of people were those who needed to talk to him and the habit had developed that if they wanted to see him, they went to his office—his "door was always open," he told me.

I suggested that we spend the 45 minutes we had doing a Gemba walk. He found this to be a wonderful and engaging experience and realized that he had to find a way to get back to this approach. He recognized he had slipped into a bad habit and needed to break it. I left him to think it over and agreed to meet again a month later.

The first thing I noticed when I arrived the next month was that his personal car park space next to the entrance had a new name sign. It now said "reserved for star of the week" and contained a motorbike. When I was shown to his office there was no queue and he was in a very different frame of mind from our previous meeting.

He had decided that to break the bad habit of always being in his office he had to create a new one. So, he moved his personal car parking space to the back of the factory and had to walk through the factory every morning and at the end of every day to get to and from his car to his office. He created a standard work routine that ensured he covered different areas on different days and that he actively engaged with people in purposeful conversations on each walk.

As a bonus, there was now a prime parking space right outside the front door. This was made available to a new person each week and allocated based on peer nominations as a recognition for something people thought should be recognized. Those nominated who did not need the space received a personal thank you and chose who they wanted to give the space to.

In this way not only was the CEO's behavior changed, but he also had a huge positive influence on the behaviors of other leaders and everyone else in the business.

In *4+1* Jones, Butterworth, Harder[15] suggest that there are three levels of maturity that organizations go through in the adoption of Gemba walks. A high-level summary of these levels is given below and these are expanded on in detail in chapter 5 of the book *4+1*.

Beginner—The organization has recognized that "Gemba" walks are important and sets a clear expectation that leaders at all levels should do them.

Intermediate—Leaders have been trained in how to do Gemba walks and they are recognized as an integral element of the System of Improvement and the System of Thinking and Behavior. Key Behavior Indicators (KBIs) are being experimented with.

Advanced—Gemba walks are regularly reviewed for effectiveness, with leaders honing their skills in shared lessons learned reviews. They are part of all leaders' standard work and are valued by both leaders and employees across the organization. KBIs are well established and in place to measure and track effectiveness.

Look, Listen, and Learn (Gemba) walks are an essential element to sustaining a culture of continuous improvement by reinforcing ideal behaviors and ensuring systems are aligned to support these behaviors. They are also an enormously powerful tool in supporting the development of leaders, managers, and colleagues at all levels. Some key takeaways to reflect on are given below:

1. Gemba walks are a process of discovery ... have an open mind and maintain curiosity throughout. It is never about validating preconceptions, but rather discovering what is going on.
2. Prepare ahead so that you concentrate on observation. Be clear on the purpose of the activity and what you want to learn.
3. Maintain respect for those working on the process, who understand it best. Engage people by directly listening and seeing. Always resist the temptation to jump in with the answers.
4. It is ok to start off at a basic level and then build up the maturity and effectiveness of the Gemba walks over time.

3.2.5 Systems Constantly Reviewed and Improved to Ensure They Drive and Support the Desired Behavior

One of the insights from the Shingo Institute teaching is that purpose and systems drive behavior. This can be in a negative or a positive way. Quite often the behaviors that a system will drive are not fully understood or even considered in the system design. Sometimes they result in the opposite of what was intended. If leaders want to instill a culture of CI, they need to regularly review and update systems to make sure they are driving and supporting the ideal behaviors.

A good place to start when reviewing a system or creating a new one is to ask two simple questions:

1. What is the purpose of this system?
2. What ideal behaviors we would like to see to deliver this purpose?

In answering the first question, a useful approach is to construct a purpose statement. An example purpose statement for a Learning and Development System is shown in Figure 3.3.

To help inform the answer to the second question, it is useful to have a workshop with those people most involved in operating the system and

▶ Learning & Development is a System that:
- Develops a learning culture & continuously develops its people aligned to the needs of the organisation

▶ By:
- Identifying Learning & Development opportunities at all levels
- Focusing on the improvement of work-competences, improvement-competencies and culture change-competences
- Undertaking pull based coaching to improve competence, performance and culture
- Ensuring that Learning & Development competences are embedded and lead to changes in behaviour and tangible results
- Recognising people for their learning achievements

▶ So that:
- The whole organisation maximises its human potential, with the members taking initiative

Figure 3.3 Learning and Development System Purpose Statement.

agree on the ideal behaviors required. These behaviors can then be used both to inform the system design and to test any proposed system.

Even when the intention of the system is well-meaning, there can often be unexpected consequences on behavior. To give an example, several years ago, the author was visiting a logistics company who were regretting introducing a new bonus system for their drivers.

> *Logistics Company Manager:* We thought the bonus system was a great idea. We wanted to reduce the amount of fuel we use, as it's one of our major costs, and we also thought that it would be a good way to support our environmental policy.
> *Author:* Sounds like a good idea.
> *Logistics Company Manager:* Yes, we thought so. Every driver now gets a bonus based on fuel efficiency. The more kilometers they do with less fuel, the higher their bonus. It's calculated in such a way that the company still saves money even with the bonus payment.
> *Author:* But it's not working?
> *Logistics Company Manager:* No. It's been a complete disaster. We just didn't realize what behavior the system would cause. You see, the simplest way to save fuel is to drive slower, which is no bad thing. But we have people driving excessively slowly, resulting in the deliveries arriving late. One of our biggest contracts is with a major supermarket and we have to pay penalty charges for every late delivery. Our penalty charges from customers have gone through the roof under the system, as we are making lots of late deliveries, but the drivers still get their bonus.
> *Author:* Oh dear.
> *Logistics Company Manager:* Exactly. And that's not all. The other way to get really good fuel consumption is to drive an empty truck. Occasionally, drivers might not be able to pick up the return load for a genuine reason and have to return to the depot with an empty truck. However, the number of instances of no return load being available has doubled since we introduced the bonus, and our truck utilization has plummeted.
> *Editor:* Sounds like it's a real headache.
> *Logistics Company Manager:* Yes, and today was the final straw. A fight broke out between two drivers because one accused the other of trying to siphon off some of his fuel so that he could get a bigger bonus. They've all gone mad!

The bonus system was reviewed and changed once the unexpected impact on behavior was understood.

As such, a key aspect of any assessment must be the review of the linkages between systems and the behaviors they are driving. Leaders need to recognize that systems will need constant adjustment and refinement to support the desired ideal behaviors and be responsive to any changes required.

3.2.6 Quickly Dealing with Behaviors that are not Desired (The Standard You Ignore Is the Standard You Set)

Sometimes it is essential to explicitly call out the behaviors that are unacceptable and not just the ideal behaviors. This is particularly important when change to an existing culture is required where behaviors have been allowed that are unacceptable. By making this public and gaining consensus, people at all levels are empowered to highlight non-conformance to the required behaviors. It helps create an environment where people feel safe to hold each other to account.

What is critical is that concerns are raised immediately, dealt with quickly, and actions fully backed by leadership. It can take a long time to build back trust, but it takes just one wrong step to destroy it again very quickly. The assessment system must be able to explore if people genuinely feel they are safe to hold colleagues and leaders to account for undesired behaviors.

These are sometimes referred to as "under the waterline" behaviors and some examples are given below from one organization that used this approach successfully.

We will not:

- tolerate or engage in behavior that leads to discrimination, harassment, or bullying;
- take credit for the contribution of others;
- turn our backs on others when they need help;
- be complacent; or
- ignore issues.

One of the most powerful ways to use this technique is to not only set some clear expectations but also encourage teams to describe in their own words the behaviors they do not want to see. This way they own the outputs and are more likely to hold each other accountable than if they are given a generic list from the wider business.

3.2.7 Role Modelling Ideal Behaviors

People believe what they see far more than what they are told. As such, it is essential that leaders personally demonstrate the desired ideal behaviors. "Do what I say, not what I do" will just not work. If we wish people to behave in a certain way, then the way leaders behave must be aligned to the ideal behavior we are looking for. Instead of *telling* people how to behave, *show* the desired behavior by demonstrating it.

This will often require a level of humility that some leaders will be uncomfortable with. One way to think about this is to realize that, as a leader, it is far more impactful to help people see how clever *they* are, not keep trying to show them how clever *you* are.

An especially useful way to role model behavior is to remove blame and accept accountability. Unfortunately, in some organizations whenever a mistake is made, the first word heard is "who" did that. This is a rich aspect to explore when undertaking a behavior-based assessment.

When people are blamed for a mistake it has an extremely negative impact on the culture of the organization. Problems are not disclosed for fear of the consequences, opportunities for improvement are not raised, and issues that could have been resolved at low cost can escalate to be extremely expensive to correct.

Also, from a people point of view, very few people come to work to deliberately do a job badly. In most cases, they want to do a good job and feel a sense of accomplishment and pride in the work they do. If they are blamed when a genuine mistake has happened, they will be defensive and understandably reluctant to accept full accountability.

If we ask "why" instead of "who" and genuinely seek to understand why the mistake has happened, we get a quite different result. An example of this is shared below.

While facilitating a leadership workshop with the senior team at a manufacturing facility it was interrupted at the coffee break on the second day by a team leader coming to the workshop and asking to see the operations manager.

A mistake had been made in setting up a machine and the resulting defective products cost over $50,000 in scrap. The team leader came to see the operations manager to explain what had just happened. They were clearly angry and upset and advised that they wanted to formally discipline the two production team members involved. Realizing there was a big opportunity to lead by example and to coach the team leader, the operations manager asked for a pause in the workshop. He then held a one-on-one coaching session with the team leader and supported the conversation the team leader then had with the operators.

Instead of summoning the team members to his office, he went to their area of work with the team leader and asked the operators if they could explain what had happened. It transpired that it was quite easy to make a mistake and insert a particular piece of equipment the wrong way round. The manager listened carefully, asked a few questions, and then apologized to the team members that the process was so poorly designed it allowed this mistake to be made. He also then asked if they would be willing to take part in a problem-solving session to work out how to change the process to make it impossible for the mistake to happen again.

The operators were overwhelmed with this response. News spread like wildfire around the whole facility and both the team members and the team leader separately thanked the manager for the way the situation was handled. All of them felt respected and spoke with pride about the way they had been treated. The incident was used as an example of the leadership behavior the company wanted to instill and resulted in a huge uplift in engagement and CI ideas. A few months later, the operations manager told me it was probably the best investment of $50,000 he had ever made.

A well-worn phrase is "the standard you ignore is the standard you set." It is worth exploring this in more detail, as it can have very negative and long-lasting consequences.

One day I was visiting a factory for a coaching session with a general manager and we agreed to start with a Gemba walk. The factory was extremely clean and well laid out, with the floors painted in color-coded areas. Walkways were clearly marked and painted green and hazardous areas painted red.

Less than one hundred meters from the main entrance from the offices to the shop floor we came across a large pallet of material completely blocking the green walkway. Without even pausing, the GM walked straight into the red zone and to get around the obstacle and carried on. I stopped and waited for him to realize he was alone.

After a few seconds, he stopped and came back to the pallet.

"What's the matter?" he says.

Realizing this was a good opportunity to practice, I asked, pointing at the pallet, "Is this supposed to be stored here?"

"Well no. I see what you mean. But we are on a tight schedule and I'll look into it later."

I asked him what he thought the consequences were of appearing to ignore the issue and not dealing with it immediately. We then had a particularly useful conversation along the lines of "the standard you ignore is the standard you set," but he also realized that what he had actually done was set a new standard. As the most senior person on site, he had ignored a standard and had in effect made it ok for others to ignore the standards. In other words, ignoring the standard was now the standard.

"OK," he said. "I get it and should definitely not have done that."

"What do you think we should do?" I enquired.

"I'm going to go and find Jim, the forklift truck driver who left it here and give him a real roasting."

"Well, we could do that but will that really solve the issue? Will we know why he left it here?"

So, we found Jim and asked him what the story was with the pallet. It turned out it had come in the wrong size and would not fit in the allocated lineside space so Jim had put it in the best place he could find. In other words, he had done the wrong thing but for the right reasons—he had tried to do his best to do a good job. After a good coaching conversation, Jim decided he would raise the issue at the afternoon stand-up meeting and get purchasing involved to discuss the situation with the supplier. We left him feeling valued and engaged in being part of the solution.

The GM felt humbled by the experience and realized that he was going to influence behavior and get much better engagement by seeking understanding rather than assuming he knew all the answers. He went to find Jim's team leader and tell him what a great job Jim was doing.

Another area where role modelling of behavior is very important is a leadership team's reactions to the CI assessment results.

Most leaders fully accept the result and realize that the benefit of the assessment is to identify opportunities and understand what actions are needed to progress. Unfortunately, we have seen several instances where leaders have been disappointed with the results of an assessment and, in a small minority of cases, the reaction is extremely negative. Leaders will denigrate the assessors, criticize the assessment process, and sometimes even complain to the CEO that they want a new team to do another assessment.

While no one likes to have their baby called ugly, and disappointment that progress is not as expected is natural—a negative reaction illustrates that the assessors were correct. Behavioral assessments at higher levels require a large amount of humility of senior management. One of the best responses to lower-than-expected feedback that I received from a CEO was "the more I learn about what we are trying to achieve the more I realize just how much more we've got to do."

I remember interviewing one senior leader on an assessment who was enthusiastically explaining some of the great CI projects that members of his team had implemented. I asked him which one he was most proud of. After a pause for reflection, he said that the project he was most proud of was the one that did not work. He had realized that with the successful projects they had celebrated and moved on without reflecting on why they had been a success. When one did not work as planned, they did a lot of analysis and reflection and learned a huge amount. They had even celebrated this learning openly and decided to reinforce that it should not be seen as a "failure" but as a real success from a learning perspective.

Role modelling of behaviors is not limited to leaders. Role modelling at team and peer-level is also important as most people will have more day-to-day contact time with their colleagues than leaders and, as such, the influence of peer role modelling should not be underestimated.

This peer role modelling can take several forms. For example:

How are mistakes dealt with within the team?
What happens if people do not keep their commitments?
What happens if people are late for agreed meetings?

Teams that agree to manage behavioral standards or norms among themselves are far more likely to work collaboratively to ensure they are followed than if they are imposed on them by other people.

One team the authors observed agreed on their own behaviors whenever they had a huddle. These included, for example, showing respect by listening and not interrupting, never criticizing any idea or suggestion, and being on time for each meeting. They had found issues with people being on time and so agreed as a team that they needed to manage this better.

Without any leadership input, they decided to put everyone's name up next to the visual management board and anyone who was late had a red cross marked next to their name. If someone got three red crosses, they had to buy a cake for everyone else on the team. Only two people had to buy cakes and then no one was late for the meeting again. This was done in a fun way that everyone bought into, but it had a positive impact on team behavior and personal accountability. This is an example of a KBI and these will be explained in more detail later in this chapter.

One important thing to look for in any assessment is how are teams using peer role modelling. Are they using KBIs (even if they do not call them this) or some other approach to support and hold each other accountable for agreed behaviors? Often a KBI might no longer be in place at the time of the assessment, as it could have served its purpose and no longer be in use. As such, it is important for assessors to seek understanding about how peer role modelling has worked.

3.2.8 Respect

One of the Shingo Cultural Enabler principles is "respect every individual." While many organizations have respect as a value, relatively few realize the full potential that embedding this principle can bring. In the authors' view, building respect into the way of thinking and behaviors is one of the most powerful things any organization can do to help embed a culture of continuous improvement.

One of the important aspects of the Shingo Principles is that it refers to the "individual" rather than a generic form of respect for everyone. This is deliberate. To genuinely show respect, we have to think about the person we are interacting with. We need to treat people as people, not numbers on payroll or a job title.

In his book, *Hearing the Voice of the Shingo Principles*,[16] Miller shares some powerful personal experiences and deep insights into this principle. He explains that the reason we should respect every individual is twofold:

1. Because every single person has intrinsic value.
2. Because every individual has unrealized potential.

I was reflecting on which organizations I had most enjoyed working for and with over the many years of my career. As I started working through the list, one thing stood out—where I had felt respected, I had felt valued. In return, I had shown respect back and put in the extra effort to go beyond what might be expected.

I was struck when reading the Nummi case study from Stanford University[17] how respect was such a driving force in their culture.

Just one of the many examples is quoted below:

> Respect was continually shown, especially to line operators, through cross-training, consideration for ergonomic impact on operators, and safety. Those in a team rotated jobs every couple of hours to share the burden and provide ergonomic relief. Team members understood that if the job were too difficult or unsafe, they would redesign the work.

What is interesting in this example is that respect is from peer to peer. Respect is not something we can show to some people and not to others, as true respect is something that must apply equally to all.

In his book *Carrots and Sticks Don't Work*[18] Paul Marciano states that respect is critical to building employee engagement. He describes his RESPECT model consisting of seven elements:

Recognition
Empowerment
Supportive Feedback
Partnering
Expectations
Consideration
Trust

He argues the case very well that reward-and-recognition systems fail to engage employees in long-term behavior change and that the only real way to achieve employee engagement is to build a culture of respect.

This is a vital consideration in the design of any assessment system. Not only must we look for evidence of respect as part of the assessment process, but also the system must be designed in such a way that it delivers and receives respect.

3.2.9 The Recruitment Process

One way to think about managing behaviors and thus culture is a process with inputs, activities, and outputs. This is illustrated below in Figure 3.4.

As with any system, it is much harder to manage effectively with poor or incorrect inputs. If we recruit people who have no sense of connection with our ideal behaviors, then it takes a lot of hard work to get this fixed. On the other hand, if we focus selection on people who demonstrate the ideal

Figure 3.4 Impact of Behavior-Focused Recruitment.

behaviors and set this requirement ahead of technical skills, then we can get some amazing results.

While many organizations look for behaviors in the recruitment process, these are often based on generic value statements or profiles of "team fit." What we are referring to here is a much more specific assessment based on the defined ideal behaviors. In addition, greater weight needs to be given to behavioral alignment rather than skillset in the final selection decision. An example of the power of this approach is illustrated in the case study below from a leading financial services organization.

Case Study—A Financial Services Organization in Western Australia

I began leading the Contact Centre in May 2019 and felt the previous leadership team had done a great job building a solid foundation for sustainable growth. I got a sense of excitement from speaking with my senior leaders, their teams, and colleagues across the division as I found they all wanted to improve and strive for better, but they lacked a vision and direction.

Understand Current State

My focus on starting my new position was to gain an understanding of how the business worked, what the culture was, what existing behaviors were evident, and what was preventing us from reaching our optimal performance.

I spent the first few weeks going on Gemba walks to look, listen, and learn how the business operated, from both observing and speaking with colleagues, managers, and leaders. I immediately noticed some instances of poor discipline and behavior, and that my leadership team was having to spend significant time each week coaching and managing those behaviors.

Further investigation of the pockets of behaviors suggested there were two fundamental causation factors:

1. Our recruitment process was heavily weighted towards "technical" ability and lacked a defined "behavioral" assessment. So, while we were recruiting colleagues with the prerequisite skills to work in a contact center, they did not necessarily fit the culture we wanted to create.
2. We had not defined a set of "expected behaviors" and, as the saying goes, "The behavior you walk past is the behavior you accept."
A priority for me was to put in place an agreed set of behaviors we would all live by, so when anyone walked through the doors of the Contact Centre—new colleagues, or existing ones—there was a clear understanding of who we were and how we operated.

Solving these two factors required a long-term strategy that involved senior leaders buying into the expected behaviors and reinforcing them through constant and consistent repetition of assessments.

We had already embedded a strong operational-excellence mindset and behaviors through our divisional assessment pathway accreditation (Bronze and Silver maturity). This tiered approach gave each division a defined pathway of capability and mindset build.

The next stage for us as a business was to move towards Gold or Shingo. At this level of maturity, leader roles shift from systems of work to systems of thinking and behaviors. We had a well-established and embedded maturity assessment but wanted to extend this to further embed ideal behaviors, which is why we decided to make behaviors an explicit element of the overall assessment.

Stage 1: Building the Expected Behaviors

It was important to create a sense of ownership over these behaviors by involving our wider team leaders in their development, so they were not seen as a set of rules developed and imposed by myself and my immediate leadership team.

It was also important to demonstrate the significance that these behaviors would hold within our culture by dedicating the appropriate time to their development. I scheduled a Team Leader Away Day out of the office and free from business as usual distractions. Six weeks prior, I had sent each attendee a copy of the book *Legacy* by James Kerr, which delves into the cultural revolution of New Zealand's revered and dominant All Blacks rugby team. It was a book that resonated with me, as it outlined the minimum behaviors required to build the foundation for a high-performing business and helped to provide a foundational understanding of what is and is not acceptable.

The away day was action-packed, with each leader reflecting on the book and what they perceived "good" looked like. By the end of the day, we had created our first version of our expected behaviors, which we then rolled out across the division. See Figure 3.5.

We also integrated an assessment of these behaviors into our onboarding process to address the first causation factor outlined earlier and to guarantee that colleagues were aligned with what we required from them both skills and cultural perspectives. See Figure 3.6 for our recruitment behavioral assessment.

We have seen a significant improvement in Key Performance Indicator (KPI) and Key Behavioral Indicator (KBI) results since the integration of these behaviors into our recruitment process. The dramatic improvement in the quality of new recruits coming into the Contact Centre has also resulted in increased capacity of leaders, who are spending less time

Figure 3.5 Expected Behaviors.

SUMMARY					
OVERALL COMMENTS					
STRENGTHS					
DEVELOPMENT AREAS/ OUSTANDING CONCERNS					
MOTIVATIONAL / VALUES ALIGNMENT	1	2	3	4	5
CUSTOMER FOCUS	1	2	3	4	5
EFFECTIVE COMMUNICATION	1	2	3	4	5
JUDGEMENT	1	2	3	4	5
DRIVE RESULTS	1	2	3	4	5
CONTINUOUS IMPROVEMENT	1	2	3	4	5
OVERALL SCORE		/30			

5	Excellent	Evidence of strength in this competency/area demonstrated consistently across all of the required behaviours
4	Good	Evidence of strength in this competency/area demonstrated across most of the required behaviours with no critical weaknesses
3	Acceptable	Evidence of strength across some of the required behaviours with minor areas of weakness or inconsistency
2	Marginal	Evidence of weakness/less than acceptable performance across a number of the required behaviours
1	Poor	Evidence of weakness/less than acceptable performance across virtually all the required behaviours. No areas of strength demonstrated

OUTCOME		
Progress	Talent Pool	Do Not Progress

Figure 3.6   Behavioral Recruitment Assessment.

managing poor behaviors and more time on quality coaching and development conversations.

The multi-step recruitment process first requires applicants to complete an online assessment. This is followed by a behavioral assessment that takes place in a group format, with activities, challenges, and role-plays. This format is designed to see how applicants interact with their peers, how they communicate with each other, and how they support each other to achieve defined outcomes.

The assessment takes a full day and is monitored and scored by a select number of leaders, as well as some existing colleagues. The applicants and potential teammates are scored by their peers, which helps create an understanding of who people want on their team. The leaders and colleagues score everyone using a defined assessment sheet, as seen in Figure 3.7.

Stage 2: Evolution of Behavioral Assessment

These behaviors set the tone and foundation of what is expected from colleagues working in the Contact Centre. It was then important to make sure they were measurable.

This was a challenge because, while our agreed behaviors were working well for our recruitment and onboarding process, it was not clear how we could measure them.

The absence of a clear link between our behaviors and our systems was also a concern. Another team leader away day was set up, this time with a focus on using the four dimensions and ten Shingo Guiding Principles to build on our behaviors.

The format of the day was broken into two parts.
The first part of the day focused on:

Addressing the question, "In our business today, what do the behaviors at each level (Leader, Manager, and Colleague) look like when each of the 10 guiding principles is NOT evident?"
The team leaders were broken up into groups of ten, each taking a guiding principle to the workshop.
Using the tools available, they were to research each principle and then present their findings to the group.
The team leaders created a visual representation of what it would look like when the behaviors were not evident.

The second part of the day explored:

What would the business look like if the behaviors were evident?
What are the beliefs of each of the 10 principles?
What are the observed expected behaviors at each level (Leader, Manager, and Colleague)?
What are the measures of these newfound KBIs?

Group Activity Evaluation Form (A)

Criteria	Team Work	Communication	Attitude	Score
Case Study:				Out of 6:
Overall Comments:				

Criteria	Below Standard = 0	Meets Standard = 1	Above Standard = 2
Team Work	• Little or no involvement • Does not include others • Lack of support • Uncooperative with both candidates & staff • Negative behaviours displays	• Inclusive of others • Gets involved • Motivated/ Enthusiasm • Helps others • Works towards end goal	• Encourages others involvement • Is engaged & participates • Motivates team • Helps others • Achieves end goal • Shows drive to achieve task • Positive behaviours displays
Communication	• No contribution • Inappropriate contributions • Does not respect others opinions • Overpowers others • Negative language	• Patient with others • Respects other views • Provides opinions/ideas respectfully • Communicates in an appropriate manner • Positive language	• Gives opinions/ideas that are well thought out • Ensures that all have a chance to speak • Confident and clear communication • Positive language
Attitude	• Late • Lack of interest/ Does not take part • Negativity towards tasks • Unwilling to learn/listen • Lack of enthusiasm	• Punctual • Participates in tasks willingly • Open to learning • Well presented • Professional towards other candidates	• Shows constant enthusiasm • Continued interest throughout tasks • Keen to learn • Approachable • Applies learnings • Motivated/ Enthusiasm

Figure 3.7 Behavioral Assessment Form Example.

We compiled these beliefs, expected behaviors, and measures (KBIs) into a "Behavioral Playbook," which we then rolled out across the division. An extract from this is shown in Figure 3.8.

The result was a set of behaviors that were linked to the 10 guiding principles and were supported by the systems we had in place.

The next challenge was the embedding of these behaviors.

I needed to make sure we continually assessed these behaviors across the division to truly drive the outcomes we wanted. We created a "Leader Playbook" (Figure 3.9) to support our leadership team and managers in reinforcing the expected behaviors. The playbook was a back-pocket set of questions linked to each principle, to which leaders and managers could refer in their one-on-ones, coaching sessions, or general Gemba walks.

Outcomes from the leaders' "look, listen, and learn (Gemba) walks" were tracked on each leader's visual management board, gaps were identified, and actions to close these gaps were then agreed on as part of our weekly operating rhythm. This transparency of our visual management board provided a level of accountability and focus that ensured the embedding of the expected behaviors across the division.

The results from this behavioral evolution have been a marked increase in Net Promotor Score and all of our KPIs across the division. Net Promoter Score increased fifty-three percent since we started working on the behaviors, which has given the team a great sense of achievement and spurred them on to continue to make tomorrow better than today—for themselves and our customers.

Our colleague engagement has also improved, with unplanned sick leave reducing from sixteen percent to six percent, and overall sentiment increasing.

As I keep saying to my team: "A continuous improvement journey is one that never ends."

We must constantly look to improve and make tomorrow better than today for our customers. The next challenge for us is to further strengthen our behaviors to guarantee that they are continually measured, tracked, and assessed.

One of my favorite quotes from the aforementioned *Legacy* book is: "When you are on the top of your game, change your game."[19] This, to me, means we cannot rest on our laurels, but continually challenge ourselves to be better.

Dimension 3: Enterprise Alignment

Belief	Think Systemically - new
	Contact Centre Belief is by having a broader lens on what may impact our customers we may be able to influence a better outcome by understanding the up and down stream impacts.
COLLEAGUE	• Understand the why of Customer and Business • Sharing Knowledge • Identifying trends/ speaking up / challenge the stratus quo • Improved behaviours and increased engagement
MANAGER	• Managers are seen to encourage colleagues to research and share internal and external information that may influence our customers. • Managers are actively working with up and down stream peers to improve the flow of information and collaboration.
LEADER	• Constantly communicate the CVP, vision, goals and connecting them to the work. • Transparency in decisions • Frequent GEMBA/ Team meetings • Provide context around external impacts • Understands the business and stakeholder needs
MEASURES	• SLAS • Positive attrition (secondment/succession opportunities) • KPIS • RiC operating effectively • Reduced sick leave • Look listen learns outside of contact centre

Figure 3.8 Extract from Behavioral Playbook.

Why Bother Assessing & Managing Behaviors? ■ 67

Manager (TL) questions for colleagues	Dimensions		
Can you tell me about a development goal you achieved this week/fortnight/month?	Cultural Enablers		
How did you deliver on our CVP this week?	Enterprise Agreement	Results	Continuous Improvement
Did you work with a team or colleague external to the Contact Centre to support a customer interaction this week? What was the outcome/what did you learn? Is there anything we can do differently next time?	Enterprise Agreement		Continuous Improvement
What were your customers saying this week? Were there any trends or CI's that we can implement on this feedback?	Results	Continuous Improvement	
Can you tell me about a CI you've come up with this past week/fortnight/month? How was this CI identified?	Continuous Improvement	Enterprise Agreement	
Did you identify any gaps or trends in our processes this week? What did you do with this?	Continuous Improvement	Enterprise Agreement	Results
How do you provide a safe working environment for your and your peers?	Continuous Improvement		Cultural Enablers
Can you tell me about a colleague you've recognised in the last week? What was it for and what action did you take?	Continuous Improvement		Cultural Enablers

Leader (CSM) questions for Managers (TL)	Dimensions		
Can you tell me how you communicate the availability of our wellbeing tools to your colleagues?	Cultural Enablers		
Can you tell me about a development goal you achieved this week/fortnight/month?	Cultural Enablers	Continuous Improvement	
How did you deliver on our CVP this week?	Enterprise Agreement	Results	Continuous Improvement
Can you tell me about a recent time where you've collaboratively worked with someone outside your direct team/Contact Centre? What was the outcome/what did you learn?	Enterprise Agreement	Continuous Improvement	Cultural Enablers
Can you tell me about a CI or process improvement you worked on or supported your colleagues with this week? What was the outcome?	Results	Continuous Improvement	
How did you encourage your colleagues this week to engage with external stakeholders outside Contact Centre for Customer Outcome/CI?	Continuous Improvement	Enterprise Agreement	Results
How do you remove roadblocks this week to enable your colleagues to seek perfection?	Continuous Improvement	Enterprise Agreement	Results
How do you provide a safe working environment for your colleagues?	Continuous Improvement		Cultural Enablers
Can you tell me about a colleague you've recognised in the last week? What was it for and what action did you take?	Continuous Improvement		Cultural Enablers

Figure 3.9 Extract from Leader Playbook.

3.3 Key Takeaways

1. There is no one magic bullet to manage behaviors—a wide range of techniques needs to be employed in a variety of ways.
2. Behaviors need to mature to new levels as understanding grows of what is possible, with ideal behaviors being constantly nurtured and reinforced.
3. New employees are key inputs to our organization's culture, and recruitment needs to be biased towards behavioral attributes over technical skills.

Notes

1. Edgar Schein, Professor, MIT Sloan School of Management.
2. Richard H. Thaler and Cass R. Sunstein, *Nudge: Improving Decisions About Health, Wealth, and Happiness* (London: Penguin Books Ltd, 2009).
3. Morgan Jones, Chris Butterworth, and Brenton Harder, *4+1: Embedding a Culture of Continuous Improvement in Financial Services* (2nd edn.) (Melbourne: Action New Thinking Ltd, 2019), pp. 12–13.
4. David Rock, *Coaching with the Brain in Mind* (New York: John Wiley & Sons Inc, 2009); *Quiet Leadership: Six Steps to Transforming Performance at Work* (New York: HarperCollins Publishers Inc, 2011).
5. Norman Doidge, *The Brain that Changes Itself* (Hawthorn: Penguin Books Australia, 2008).
6. Kristen Hansen, *TRACTION: The Neuroscience of Leadership and Performance*, (2017). www.enhansenperformance.com.au/traction.
7. Jones, Butterworth, and Harder, *4+1*, 43.
8. David Rock, "SCARF Model," *NeuroLeadership Journal* 1 (2008). https://qrisnetwork.org/sites/default/files/materials/SCARF%20A%20Brain-based%20Model%20for%20Collaborating%20with%20and%20Influencing%20Others.pdf
9. Hines and Butterworth, *Essence of Excellence*, 148.
10. www.octanner.com/au/ (2015, 2017).
11. Dominic Bria, "Best Ways for Manufacturers to Boost Employee Engagement," accessed March 21, 2018, https://shingo.org/best-ways-for-manufacturers-to-boost-employee-engagement/.
12. Types of Gemba walk themes developed from an original concept by David Mann.
13. Michael Bremer, *How To Do A Gemba Walk - A Leader's Guide* (The Cumberland Group, 2016).

14 Bremer, "How to do a Gemba Walk The Cumberland Group," *Essence of Excellence* (2016), 5.
15 Jones and Butterworth, *4+1*, 137–141.
16 Robert Derald Miller, *Hearing the Voice of the Shingo Principles* (Portland: Taylor & Francis Inc, 2018), 82.
17 (ref CASE: HR-11 DATE: 12/2/1998 (REV'D. 11/19/04)), 3.
18 Paul L. Marciano, *Carrots and Sticks Don't Work: Build a Culture of Employee Engagement with the Principles of Respect* (New York: McGraw-Hill Education, 2018).
19 James Kerr, *Legacy* (London: Little Brown Book Group, 2013).

Chapter 4

Why Bother Designing Your Own Strategic Level Behavioral Assessment System?

Chapter Summary

Each organization has a unique context with a different start point, different culture, different business language, geographic complexities, and organization structure. As such, a lot of thought has been given to the approach, design, and content of the assessment and customization to local requirements is critical to success. Designing your own assessment system will support the long-term sustainability of your CI culture. This chapter explains a detailed step-by-step guide showing how to design and build your own CI maturity assessment system.

Traditionally assessments have focused on the tools—the "whats;" for example, a visual management board or a workplace organization effectiveness assessment such as 5S. Even more advanced assessments focus on the effectiveness of the process or system that the tool is a part of. For example, how effectively does the visual management board integrate with the strategy deployment or Hoshin Kanri system? However, the most effective assessments not only look at both of these but also, most importantly, focus on looking for evidence of the ideal behaviors needed for long-term sustainability.

Another way to think about this is that by assessing a particular tool we are assessing the "what" and by assessing the system that the tool is a part of we are assessing the "how." By assessing the behaviors, we are assessing the "why."

If we take the example of a visual management board, the type of questions would be:

Tool-based—Do you have a visual management board?
System-based—How do you use your visual management board?
Behavior-based—Why do you have a visual management board?

A good place to start in designing your own CI assessment system is to understand what currently exists. Many organizations will already have various forms of assessments, or more likely audits, in place. Understanding how these are used and their purpose will give valuable insights into how to build on these and how to make it clear that the assessment system is different from audits and will add real value.

Once this understanding has been gained, a useful next step in designing your assessment system is to get agreement on its purpose. Once this has been agreed on, you can then explore what ideal behaviors you want to encourage and support through the application of the system.

With the purpose and initial ideal behaviors as key inputs, you are ready to start the more detailed design of the assessment system. The easiest way to do this is to go through a series of questions. Consider using workshops with key stakeholders to formulate consensus on the key questions.

One way to think about this is as a Plan, Do, Check, Act cycle:

Plan—designing an assessment
Plan—training assessors
Do—doing a pilot/learning by doing assessment
Check—review and refine approach—what did we learn about how we did the assessment
Act—repeat the cycle

It will never be perfect and it is essential to be open to feedback and to take a collaborative approach to involve internal customers in the development and ongoing improvement of the system.

4.1 Clarity on Purpose

Always start with the question "Why do we need an assessment system?" As we stated in the introduction, the authors' view is that we need an assessment system to learn where we are now and what the key steps are that we need to take to get even better. However, this is a high-level generic why and each organization should work through the answer to this question for themselves.

It is always useful to explicitly state what the purpose is not. For example, it is not:

- another audit
- a way to catch people out
- a way to justify the existence of the CI team

Having made this clear, we can now specify what we want the assessment system to achieve. This is very context-specific but some generic points would typically be:

- Understand what is working well in our approach to CI that we can build on
- Understand what we need to improve further in our approach
- Help leadership teams to understand that key activities are needed to take them to the next level
- Provide opportunities for recognition
- Create learning opportunities for everyone involved in an assessment
- Help the organization to understand and measure progress
- Help embed the ideal behaviors that will deliver ideal results

4.1.1 Examples of Behaviors We Want the Assessment System to Encourage and Support

It is an extremely useful activity to workshop both the purpose and the ideal behaviors and involve the potential customers as soon as possible in the process. Some example behaviors we want as an outcome of a good assessment system are given below.

Collaboration, for example, sharing of what has not worked and lessons learned from it. Too often the emphasis is on sharing of "best practice."

While this is also a desired behavior, over-emphasis on this can lead to driving rivalry and not-invented-here attitudes. Another important aspect of this is that more lessons are learned when something does not work as planned than when something does work as expected.

A multi-site organization that one of the authors worked with held six-monthly reviews with CI champions and site leaders to review assessment results and share lessons learned. At the first meeting, there were a lot of great success stories and recognition all around. However, it was apparent that the learning had been limited, as people were more focused on showing the CEO and senior team how well they were progressing than sharing real lessons learned.

For the subsequent session, the agenda was changed. Each site was asked to share their top three things that had not worked as well as they had hoped, what they had learned from this, and what actions they had taken to overcome the issues. Each team was also asked to present on what they saw as their biggest challenge in the next three to six months. This was then followed by multi-site groups problem solving on the challenges.

The feedback from this approach was far more positive than the first session and changed the whole way that the assessment process was perceived. This format was adopted for all future sessions.

Other examples of desired behaviors:

Leaders are advocates for the assessment system and actively promote its value.
Leaders pull for the assessment to be undertaken in their area.
Leaders respond positively to assessment findings and proactively develop and implement action plans based on feedback.
Leaders are constantly striving to embed ideal behaviors.
Leaders constantly seek to refine their systems to drive ideal behaviors.
Everyone proactively seeks out and implements opportunities to improve.
There is constant recognition and reinforcement of ideal behaviors at every opportunity.

4.1.2 Examples of Behaviors We Do Not Want the Assessment System to Drive

It can also be useful to explicitly call out the behaviors we do not want to see. An inappropriately designed system can lead to people:

Hiding issues
Rivalry
Excessive competition, where "winning" is seen as getting a higher score than anyone else at any cost
Chasing the number or score in the assessment rather than focusing on the learning and opportunities
Complacency. People believing they are already at their best and therefore do not need to do any more improvement
"Gaming the system." For example, training interviewees how to answer assessors' questions (this never works, as a trained assessor can spot this very quickly)

4.2 Designing the Assessment System

Once the purpose statement has been drafted, the high-level ideal behaviors have been formulated, and the sponsor has been fully engaged, then the detailed design can start. The order below should be viewed as iterative, as answers to questions will undoubtedly develop along with the development of the system. For example, while the agreed purpose will inform who the customers are, having a clear answer to this question will also help determine if the purpose is correct.

Below are listed the key questions to explore and gain consensus on. Each of these will be explored in more detail further in the chapter.

1. Who are the customers of the assessment system?
2. Which CI model will be used as the framework?
3. What is the scope of the assessment?
4. What are the different maturity levels?
5. What are the descriptors for each level including behaviors?
6. What are the pre-requisites for each level?
7. What is the recognition/award system?
8. What are the targets?
9. What are the routines and standard work for the assessment system?
10. What is the communication plan?
11. Planning and scoping an assessment.

76 ■ *Why Bother?*

4.3 The Key Questions

4.3.1 Who Are the Customers of the Assessment System?

This question needs a lot of thought and consultation. One approach is to draft the purpose statement and then compile a stakeholder analysis based on this. For example, based on the purpose statement, who are the potential stakeholders and customers? At this early stage, it is useful to approach the identified customers and gain consensus on the purpose. Remember, we need to build a system that both demonstrates respect and is respected.

Potential customers could be some or all of the below:

- the leadership team
- the sponsor
- the heads of functions and departments
- all employees
- suppliers
- the external customer
- a particular leadership group or site

Having identified the customers, next is to discuss with them what they would value most from the system. One way to do this is through "voice of the customer" interviews and to agree to a high-level "customer value proposition." In an ideal world, the purpose statement could be the same as the agreed value proposition. At the very least, they need to be closely aligned. The customers could evolve and expand over time—we do not have to try to solve everything at once.

4.3.2 Which CI Model Will be Used as the Framework?

Most organizations considering adopting an assessment system will already have an agreed CI framework to which they are operating. Typical examples could be:

- the Shingo Model (see Figure 2.6)
- the Lean business model (see the Panalpina case)
- an internally developed model, for example, the X company business system

The particular model or framework is important and some of the key critical features are listed below.

Must be well understood and owned by the senior team.
Must be easy to explain and communicate to everyone in the organization.
Must have a strong focus on organization culture and behaviors.
It must be the leadership team's model, not the CI team's model.

Very rarely it is appropriate to cut and paste a standard model without any local adaptation. Every organization is unique. It has its own context, systems, business language, and culture. Therefore, the best assessment system will be customized to the organization and based on an improvement model that the organization has designed to work for them.

That said, it is not necessary to start with a blank piece of paper. We highly recommend using the Shingo model as the basis to inform your own approach and, where necessary, customizing this to your unique requirements.

4.3.3 What Is the Scope of the Assessment?

It is important to define the scope of the assessment system. It may be appropriate to start small and then broaden the scope based on lessons learned. The best systems can be adapted easily to cover multiple scopes. Some key considerations are:

4.3.3.1 Will the Assessment Be Focused at the Site Level?

This may be more complex than a simple geographic decision. For example, several teams based at multiple locations may share a common management team. Some organizations may have huge sites with several thousand people. The key factor to take into account here is that the assessment result must be meaningful in terms of recognition and identifying the opportunities. Too big a group and the feedback will become too generic and of little value. Too small and it risks being more based on individuals rather than system-driven. If a site is very large, then we recommend breaking it up into appropriate business units each with around 200–300 people and directing the feedback and action planning to the leadership group of each business unit.

4.3.3.2 How Will Functions That Are Not Site-Based Be Assessed?

This is a common concern about assessment systems. Functions covering, for example, people or finance or IT often have complex locations and dual reporting structures. The best place to start is understanding the organizational structure. Function teams based at a site to support the site but with a reporting line to their central function can be included in the site-based assessment. Centralized functional teams can be assessed in the same way as a site. One of the advantages of focusing on behaviors is that, in terms of the assessment, it does not really matter what discipline people work in.

4.3.3.3 Will the Assessment Need to Be Provided in Several Languages?

To do an effective assessment, it needs to be conducted in the local language. It is possible but not ideal to do these using translators, and in the early stages, this may be the only option. Over time, however, there needs to be a development program that trains local language speakers in the assessment process.

4.3.3.4 How Will the Assessment Work at the Local Level?

Ideally, the assessment needs to be designed in a way that allows local coaches and leaders to do ongoing regular "temperature checks" or "mini-assessments" in their areas. There should be no big surprises in assessment results. We are not trying to catch people out but to help them celebrate success and go on to the next level. It can be useful to create summarized versions of the assessment that can be used locally for self-assessment checks. An example of one of these is shown in Figure 4.1.

4.3.3.5 What Is the Minimum Group Size That Can Qualify for the Assessment?

There are several reasons why there needs to be a maximum and a minimum group size that can qualify for a formal assessment. This is explored in more detail in Section 9.2 later in this chapter.

Principle	Behaviours	Happens in a few teams	Well established in several teams	Well established in all teams and always improving	Embedded as a way of life in all areas
Create Value for the Customer	We listen to our customers and act on what they say				
	We have an endless search for new ways to give our customers and team something better.				
Create Constancy of Purpose	We take actions every day to make our colleagues and our customers lives a little better				
	At our visual management board and huddles we regularly discuss measures that matter to the customer				
Think Systemically	When we think of a solution, we think of how it affects different teams around us, how it works seamlessly for stores, and how it helps our customers.				

Figure 4.1 Example Simple Summary Assessment.

Principle	Behaviours	Happens in a few teams	Well established in several teams	Well established in all teams and always improving	Embedded as a way of life in all areas
	We work as one company, working within one operating framework.				
Flow and Pull Value	We use 5S to create very effective workplace organisation with a place for everything and everything in it's place				
	We constantly seek to remove or reduce any of the 8 wastes we see and discuss these at least once a week				
Assure Quality at Source	Our Standard Operating Procedures are easy to follow and easy to update when we have improvements				
	Everyone takes personal accountability to try and achieve "right first time" in all our actions				

Figure 4.1 Cont.

Principle	Behaviours	Happens in a few teams	Well established in several teams	Well established in all teams and always improving	Embedded as a way of life in all areas
Focus on Process	We regularly review our processes to see how we can make them better for the customer and simpler for each other				
	When a mistake happens we review how to improve the process not blame the person				
Embrace Scientific Thinking	We use the 5 why's problem solving approach whenever we have a problem				
	Leaders coach us in problem solving tools				
Seek Perfection	When we see a customer experience that could be better, we go above and beyond to improve it.				
	We do what's right - we treat our business like our own, we use our initiative and we are proactive.				

Figure 4.1 Cont.

Principle	Behaviours	Happens in a few teams	Well established in several teams	Well established in all teams and always improving	Embedded as a way of life in all areas
Lead with Humility	We are encouraged and supported to think for ourselves, if we believe it will help our customers.				
	Leaders value our opinions and encourage contributions from all team members				
Respect Every Individual	We rise to challenges together, listen to each other and build on each other's work.				
	We regularly see senior leaders attending our huddles and interacting positively with team members in the store				

Figure 4.1 Cont.

4.3.3.6 Will the Assessment Cover Suppliers?

This is very rarely the best place to start but organizations are increasingly becoming more dependent on their supply chain than their internal processes. For example, several high-profile food brands do not own any facilities of their own. Nevertheless, a behavioral assessment system for their organization is still very useful. One large retailer uses a behavioral-based assessment system to grade all their suppliers and awards future contracts based on the results. There may already be an existing supplier-grading structure in place and the opportunity is then to integrate the behavioral assessment system into this. Context is key but it is usually best to start internally and progress to external assessments once the organization has reached a minimum level of maturity itself.

4.3.4 What Are the Different Maturity Levels?

There is a wide range of options here. The most common is to have several stages ranging from three to five. The exact words differ but the stages are usually very similar to:

not yet started or individually driven
have some activity in local areas based on an agreed standard
well established in many areas
well established and continuously improving
a way of life—it is just the way we do things

There are several academic viewpoints on levels of maturity across organizations. Some of the more widely known include *High-Involvement Through Continuous Improvement*;[1] *Theory in Practice: Increasing Professional Effectiveness*;[2] and *The Smarter Organization: How to Build a Business That Learns and Adapts to Marketplace Need*.[3] All of these give particularly useful insights on characteristics that can be observed in organizations that develop an embedded CI culture. These perspectives are not mutually exclusive but rather reinforce each other and provide an extremely useful way to view progress on the journey. These are summarized in Figure 4.2.

Every organization must start somewhere and, in terms of assessment, the starting point is less important than the action plan it generates to take the organization to the next level. The Shingo Institute awards recognition at three levels—Bronze and Silver medallions and The Shingo Prize. These are summarized as:

Bronze—Organizations in the earlier stages of cultural transformation with a primary focus on tools and improvement.

84 ■ Why Bother?

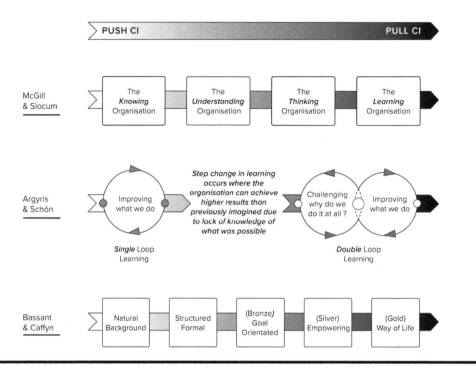

Figure 4.2 Summary of Phases of Maturity.

Silver—Organizations maturing on the journey with a primary focus on tools and systems for improvement.

The Prize (Gold)—A worldwide recognized symbol of an organization's successful establishment of a culture anchored on principles of organizational excellence.

These levels are roughly equivalent to Level 3 (Bronze), Level 4 (Silver), and Level 5 (The Shingo Prize) shown in Figure 4.2.

Some organizations choose not to have assessments at all five levels and instead not start formal assessments until they have been on the journey for some time and feel that they are close to achieving Bronze/Level 3. This can be a valid choice and depends on context, but it is usually a result of misconceptions about the purpose of the assessment system.

It is quite common for organizations to say that an assessment is a waste of time if they have not really started yet as they will just be told they are at Level 1. Unfortunately, this can be true if the assessment is structured just to give a score. However, if it is structured instead to identify what is working well and can be built on, and what the key opportunities are to develop, it will be very valuable regardless of the actual numerical result.

Other organizations feel it is important to understand the baseline starting point, no matter what it is, so that they can track progress and the effectiveness of the CI journey. If designed and used correctly, the assessment system will give valued guidance and recognition and provide useful insights regardless of the maturity level.

It is important to not let perfection stop improvement and this maxim applies equally to designing the CI assessment system. It will never be perfect. Instead, it needs to incorporate PDCA thinking and constantly evolve. Some organizations spend many years refining the detail of every level only to find that by the time any area has neared Stage 5 their understanding and perspective of what is needed has completely changed. Instead, we recommend keeping the detail at a high level for the more mature levels and then developing the details for these collaboratively with the parts of the organization that want to actively pursue them.

4.3.5 What Are the Descriptors for Each Level Including Behaviors?

As can be seen from the example below, it is often not the case that completely different requirements are needed at each maturity stage. Instead, it is more about the extent (how widespread), the depth (how far up and down the practice is established), and the length of time the element has been in place. For example, Stage 5 is unlikely to be achieved in less than three years and often takes at least five.

An example of the kinds of statements that describe what can be expected to be in place at different stages of maturity is shown in Figure 4.3.

	Not Started	Elements in place in some areas	Standard implemented in several areas	Well established and being improved	Fully integrated into Day to Day routines
	1	2	3	4	5
Understand Voice of Customer	Voice of Customer concept not understood.	Voice of the Customer concept understood but not clearly defined and calibrated with the customer.	Most people can identify their internal and external customers. Several teams have customer focused metrics based on VOC activity that drives improvement.	Most people and teams can identify their internal and external customers. Most teams have customer focused metrics based on VOC activity that drives improvement. Formal review routines are established with key customers.	Everyone can identify their internal and external customers. All teams have customer focused metrics based on VOC that drive improvement. Customer feedback sought and collaboratively acted upon.

Figure 4.3 Example Statements at Maturity Levels for a Single Element.

The elements will be based on your chosen CI model. For example, the Shingo Institute assessment is based on four high-level dimensions of:

1. Results
2. Enterprise alignment
3. Continuous improvement
4. Cultural enablers

The trap to avoid is to not make the levels of maturity about CI tools, but instead about the behaviors that the organization wants to see in place. As discussed in Chapter 3, ideal results require ideal behaviors. As such, we do not want a measurement of maturity, for example, to be based on the number of teams that have visual management boards or that are applying 5S. Instead, we want to assess the ideal behaviors.

At a high level, one way to think about this is that at the lower levels of maturity people will generally understand "what" is required; for example, we need to have a visual management board. Evidence that maturity is increasing is demonstrated by them understanding how something is used and how it fits into a wider system (e.g., how a visual board helps link the team to the customer and the organization's purpose). As maturity increases, they will understand why something is used. For example, a visual management board helps us to set priorities based on customer and business needs and track progress against our goals that support those needs. These phases of understanding tend to apply at all levels across an organization.

One way to think about behaviors in terms of CI assessment levels is that there are six broad levels:

Disruptive: People actively seek to sabotage improvement activity. This is rare but unfortunately can sometimes be seen in a small number of individuals and needs addressing quickly.
Negative: People are disengaged, openly do not want to be involved in CI, and see it as extra work they are being burdened with. (This is not a judgment of them and is most likely to be the result of the way they are treated or the systems they have been given to do their job.)
Passive: People do not want to make changes and are happy with the way things are. They support the status quo and can often be typified with an "if it's not broke, don't fix it" attitude. This is often a result of previous bad experiences.

Reactive: People are happy to take part when asked. They will contribute when encouraged to but do not take the initiative.

Proactive: People seek out improvement opportunities, share learning, seek knowledge, and constantly look at how they can improve themselves and the systems they use.

Passionate: People are proactive but also advocates for improvement and seek to implement, teach, and coach at every opportunity. They are not "evangelical zealots," but their enthusiasm is infectious.

These high-level categories can be used to inform the expected behavior at each level. Behavior is rarely going to be uniform until the higher stages of maturity are reached and it is likely there will be wide variations within each team. The behavioral level is assessed in the answers people give and the way they are observed to behave in meetings and interactions with colleagues. This will always have a level of subjectivity and comes with a health warning. We are looking for the way most people behave and are also assessing whether this is accidental or part of a deliberate effort to manage behaviors in a certain way.

While the broad categories above can be useful in terms of informing our definitions, an effective behavioral assessment can only be conducted if ideal behaviors have been defined and deployed. We need to have specific behaviors developed that people can aspire to and that can be assessed. However, it is important not to try to centrally define generic behaviors for each team as these will have no real belief for the team. Instead, set guidance and coaching and lead by example in the behaviors required. Ideally, as maturity increases, teams can be assessed against the behaviors they themselves have set. For more details on behavioral deployment see *The Essence of Excellence*.[4]

A very useful high-level summary of behavioral assessment is provided by the Shingo Institute and is reproduced in Table 4.1.[5]

4.3.6 What Are the Pre-requisites for Each Level?

Advancing through the progressive levels of maturity in the assessment system should not stand in isolation from other business criteria. For example, the whole credibility of the assessment system could be called into question if a team is awarded a high level of CI maturity but is failing to meet any of their financial targets.

Table 4.1 Behavioral Assessment Scale

Lenses		Level 1: 0–20%	Level 2: 21–40%	Level 3: 41–60%	Level 4: 61–80%	Level 5: 81–100%
Role		**Executives** are mostly focused on fire-fighting and largely absent from improvement efforts.	**Executives** are aware of others' initiatives to improve but largely uninvolved.	**Executives** set the direction for improvement and support the efforts of others.	**Executives** are involved in improvement efforts and support the alignment of principles of operational excellence with systems.	**Executives** are focused on ensuring the principles of organizational excellence are driven deeply into the culture and regularly assessed for improvement.
		Managers are oriented toward getting results "at all costs."	**Managers** mostly look to specialists to create improvement through project orientation.	**Managers** are involved in developing systems and helping others use tools effectively.	**Managers** focus on driving behaviors through the design of systems.	**Managers** are primarily focused on continuously improving systems to drive behavior more closely aligned with principles of organizational excellence.
		Team members focus on doing their jobs and are largely treated as an expense.	**Team members** are occasionally asked to participate on an improvement team usually led by someone outside their natural work team.	**Team members** are trained and participate in improvement projects.	**Team members** are involved every day in using tools to drive continuous improvement in their own areas of responsibility.	**Team members** understand principles, "the why" behind the tools, and are leaders for improving not only their own work systems but also others within their value stream.
Frequency		Infrequent Rare	Event-based Irregular	Frequent Common	Consistent Predominant	Constant Uniform
Duration		Initiated Undeveloped	Experimental Formative	Repeatable Predictable	Established Stable	Culturally Ingrained Mature
Intensity		Apathetic Indifferent	Apparent Individual Commitment	Moderate Local Commitment	Persistent Wide Commitment	Tenacious Full Commitment
Scope		Isolated Point Solution	Silos Internal Value Stream	Predominantly Operations Functional Value Stream	Multiple Business Processes Integrated Value Stream	Enterprise-wide Extended Value Stream

© The Shingo Institute

The most effective approach is to incorporate key business requirements as pre-requisites for attaining a particular level. For example, a team or site must achieve a minimum performance in financial requirements before they can qualify to apply to be assessed at a higher level.

Three areas we recommend as a minimum to use as pre-requisites are:

1. Financial performance—for example, must be beating or at least achieving the agreed budget.
2. Safety must be meeting all safety standards.
3. Must be meeting the agreed levels on employee engagement measurements.

The targets for Stage 5 are typically higher than for Stage 4.

Other areas that may be worth considering are formal Voice of the Customer results or process maturity levels, if this is a tool used in the business. Some organizations also require a minimum percentage of leaders and the wider business achieving sign-off on CI skills to an agreed standard. Also built into this is a published CI skills development plan that seeks to ensure ongoing sustainability and application of skills learned.

There may well be other criteria you wish to include. This is an opportunity to collaborate with other teams across the business and build their requirements into maturity levels.

A useful approach for teams seeking to be assessed at Stage 4 or Stage 5 is to get them to complete an "achievement log" or communication document on why they feel they are ready to be assessed at the level. This is a fantastic way for them to celebrate success and recognize people. The most common feedback we hear when teams do this is, "We had forgotten just how much we have done."

This also puts the onus on teams to pull for the higher-level assessments rather than the assessment team having to push them. Leaders seeking recognition at this level and pulling for assessments are far more effective at driving change and supporting sustainability.

One way to look at this is that the CI assessment is an umbrella assessment that builds on existing requirements and is a tool for integration across the organization. Targets can be set for existing requirements, such as process maturity level as a pre-requisite for a particular CI maturity level.

4.3.7 What Is the Recognition/Reward System?

Recognition is a key driver of behavior.[6] If we want the assessment to be valued and to drive behavior, then recognition of achievement is key. We recommend avoiding financial rewards as recognition and instead focus on public acknowledgment and building pride in the results.

It is important that the recognition celebrates improvement rather than just an absolute number. Recognition should not be about the score—leagues tables comparing one group's score against another will just drive undesired behaviors. We want people to collaborate to help others to improve, not compete to achieve a number. Therefore, try to build recognition mechanisms that emphasize collaboration, learning, and achieving progress.

One organization, for example, issues formal certificates signed and presented by the CEO to each team that completes the assessment. Teams that achieve progression in their results see a higher proportion of internal promotions (a deliberate policy) and are rewarded with case studies in newsletters and visits from other teams. Also, they get the opportunity to take part in external benchmarking activities. They become known as a "Silver Team" and are rightly proud of the accolade. Consequently, that teams' environment is viewed as a "great place to work" and attracts even more talent—this is essential, as they also face the challenge of other teams actively trying to recruit their great people.

While the assessment system should have formal public recognition activities built-in, it should also encourage the development of local-level recognition, giving freedom to local leaders to recognize achievements on the CI journey and as a direct result of the assessment outcomes.

One golden rule of recognition is that you can never recognize too much but make sure you recognize the right things.

Many organizations value external recognition, and challenging for a Shingo Prize is a great example of this. Not only can be it be used to recognize fantastic efforts across an organization, but also it provides excellent insight into further opportunities. Indeed, for many challenging organizations, the opportunity to learn is more important than the potential award. More details on the challenge requirements can be found on the Shingo website www.shingo.org.

4.3.8 What Are the Targets?

Setting targets can be a tricky area. Done incorrectly, it drives undesired behaviors and results in leaders chasing a number rather than truly implementing a CI culture. Like many things in CI, it is not *what* we do but *how* we do it that is critical.

One approach that we have seen work well at a global business level was for the executive team to set a minimum level that business units were expected to achieve within two years. This was set at Level 3 of their five-level model. This had the benefit of creating a pull from the early adopters who were ready and willing to start and their progress then meant other business units saw the need to get involved.

Not everyone achieved the goal but the few who did not stand out realized that if they wanted to be seen as successful in the business this was something they needed to do. Consequently, they actively sought to achieve the minimum standard in year three.

Interestingly, there was a deliberate policy decision by the executive team not to set a target date for levels four and five. Instead, it was made clear that it was an aspiration to get to these levels and leaders could choose if and when they thought they were ready to try for it. This created a lot of healthy competition and several business units achieved Level 4 in year three and subsequently went on to work towards Level 5, which was rewarded with external recognition.

The key to successfully setting targets for CI assessment is to ensure that the purpose is fully understood and constantly reinforced. The purpose is not to achieve a certain score—if it becomes this, then it just evolves into another tick-box exercise and a waste of time. It is important to constantly communicate and remind people that the purpose is to recognize what is working well and identify opportunities for further improvement. The score and achieving a particular target are purely an indication of the progress, not the purpose.

4.3.9 What Are the Routines and Standard Work for the Assessment System?

There are several areas that need to be designed and agreed on. These will change over time as lessons are learned and routines are improved, but in the initial planning it is vital to agree on some key things:

1. Frequency of assessments
2. What size of team will be assessed
3. Duration of an assessment
4. Who to interview
5. Observations outside of interviews
6. Number of assessors to undertake an assessment
7. Documentation and feedback of the results
8. Lessons learned reviews and system improvement

4.3.9.1 Frequency of Assessments

There are numerous choices to make around the frequency of the assessments. One option is to set the frequency based on fixed timings—for example, every six months or once a year. Alternatively, the trigger for an assessment could be timing around particular milestones. While both approaches have advantages and disadvantages, what we have found is that a combination of both works well.

If it is agreed that a baseline is useful (and we recommend this), then this can be timed to align with an agreed starting date for a team and used to inform their initial implementation plans. Future assessments could then be timed to align with expected milestones but should not be more than twelve months apart. However, assessments in the early phases are useful on a more frequent basis and every six months in this first two years is not uncommon.

I remember one leader actively encouraging assessments on a more frequent basis and when I asked why he was so keen to do this, his response was, "It helps keep us honest. I do not want us kidding ourselves about our progress, but also the external perspective is a big motivator to guarantee we do it correctly." They went on to achieve very high levels of CI maturity.

Another approach is to allow for customer pull by setting a long-term aspirational goal (e.g., achieve Stage 3 in the next two years) and allowing teams to pull for assessments when they feel ready or decide they need guidance.

4.3.9.2 Size of Team to be Assessed

In some organizations, this may be a straightforward decision and can be set at the site level. However, this is often not the case and it can become a complex problem when dealing with multi-national matrix organizations. In other organizations, a site may consist of several thousand people effectively working in stand-alone business units.

A rough rule of thumb is that a site of more than two hundred to three hundred people will need to be broken down into smaller units for the assessment results to be meaningful. Any bigger than this and the results risk being so generic that they become an "average elephant" that no one can relate to. When this happens, it is difficult for the feedback to be turned into meaningful actions. As such, it may be necessary to break down large groups into more meaningful sub-groups.

Additional criteria other than size can be used to aid this decision, such as which groups have a defined leadership team or identifiable customers. There will always be exceptions to whatever criteria are decided and it is important to recognize this and be prepared to adapt the system to local needs as required.

One thing to tightly control is the minimum size of the team that can be assessed. In several organizations, we have seen local leaders demanding to be assessed separately because "their team is different" from the larger business unit or function being assessed and they want an individual score for their team. This illustrates a real misunderstanding about the purpose and such demands should be seen as a lead indicator that something has gone wrong in the communication.

It may well be that they feel they are at a more "advanced" level than other teams and will not get the recognition they feel they or the team deserve. However, rather than give them a separate assessment, we recommend shining a light on their team as part of the larger process and calling out (recognizing) in the overall feedback what they are doing well. Instead of seeking a separate assessment, they should be encouraged to collaborate and support other teams.

One criterion that can be useful is to set a minimum level in the reporting hierarchy. For example, a team must report to management level X to qualify for an assessment. Assessing a small sub-team at a high maturity level can be difficult to sustain, as a relatively small personnel change can have negative flow-on consequences. Also, the sub-team will have much less influence over their own direction and risk being derailed by things outside of their control such as changes in priorities or unexpected budget cuts. Providing an assessment for a sub-team also risks highlighting undesired behaviors and could undermine the credibility of the assessment system. At a practical level, it will also put a lot of pressure on the finite capacity of the assessment team and risks the assessment team being overwhelmed with demand for assessments.

While it is important not to go too granular with sub-teams that have a limited span of control over their direction, this is not the case with stand-alone business units. Where there is a fully functional business unit with its own leadership team and strategy, these can be assessed regardless of size and are often able to make rapid progress.

4.3.9.3 Duration of the Assessment

To gain a good understanding of the current level of maturity of the organization and refining the feedback to the vital few areas takes time and reflection. There is a lot of work to do and this is often underestimated. While any skilled and experienced assessor could undertake an assessment on their own, we do not recommend this. The scale of the area being assessed may make this impossible in any case, but even if a lone assessor is feasible we recommend always having at least two assessors.

Ideally, always take the opportunity to include a third in an observation or learning capacity. This could be a leader from another area who has no intention of becoming an assessor but wants to understand how the process works and find out what good looks like. This should be actively encouraged as it is an immensely powerful learning and development experience.

The reason we recommend at least two people is that there is always a level of subjectivity in any assessment. We need to be able to cross-reference, debate, and validate what we have seen to reduce the risk of missing something important or focusing on the wrong thing. Also, the calibration of results from different perspectives makes the output much stronger.

Typically, an assessment will take place over two to four days, depending on the size and scope of the business unit or team being assessed. To give one example, a Shingo Institute assessment for a business unit will use a team of at least five people over three days.

4.3.9.4 Who to Interview

Planning the interviews is critical and they should not consist of a random number of people being selected. It is important to have representatives from all areas of the organization and across all levels. Generally, the number of people interviewed from each area and level should be a representative proportion of the number of people in each of those areas and levels. For example, there might be eight senior leaders and two hundred colleagues. Interviewing all eight leaders and only four colleagues will not give a representative picture. For most groups of between one hundred and three hundred people, experience has shown us that around twenty-five people in total will be enough. Doing more than this does not really highlight any further insights and doing too few will give a skewed result. Typically, forty-five minutes to an hour is plenty of time for leaders and managers, and most colleague interviews can be completed in thirty minutes.

Always have a detailed interview plan agreed on before starting the assessment and review the list of proposed interviewees to ensure the right mix and spread. It is essential to create standard work around this and share it in advance with the teams due to be assessed.

4.3.9.5 Observations Outside of Interviews

Interviews alone are insufficient for any assessment. Assessments should also include a range of other activities, such as:

- Gemba walks by the assessors (ideally accompanying a leader or manager as an observer)
- Sitting in on key meetings (these can be recorded and watched offline if needed)
- Just watching various workplace areas and offices to observe the type of interactions and behaviors

4.3.9.6 Number of Assessors to Undertake an Assessment

It is important to map out the expected number of assessments, their frequency, and their scope to understand the resource requirements needed to support the assessment system.

The system design needs to consider where these assessors will come from and how will they be trained. The team should be a mix of full-time assessors and senior people from across the business who are seconded to the team to undertake specific assessments. Larger teams will typically have two or even three levels of assessors, such as lead assessors, assessors, and trainee assessors. More detail on this is given in Chapter 8.

It is a fantastic opportunity for people to get a real insight across the whole business. I once had a new finance director spend three days of her second week shadowing the assessment. Her feedback was that it was the best induction she had ever had.

The assessment system should also be designed to allow for quick "temperature checks" or mini local assessments that can be conducted by local CI support and/or leaders and managers. The more they understand how the assessment works, the more they will value it and the greater chance of it being valued as something useful rather than an audit by someone else. Ideally, the central team's assessment should be confirmation of what the local leadership team already expects in terms of the current state, but with expert guidance on the way forward and external recognition of what has been achieved.

4.3.9.7 Documentation and Feedback of the Results

The key to feedback on the assessment results is that less is more. Design a feedback format that focuses on the vital few. Try and build on existing plans (they may need refinement) wherever possible and make sure the why behind any opportunity is clearly understood.

Typically, a feedback report will consist of:

- a thank you slide
- one slide on each high-level element assessed
    - what is working well (strengths)
    - what could be even better (opportunities)
    - a summary slide
- key themes and recommendations
- ideally use quotations (not attributed, unless agreed) and specific observations to back up findings and photographs
- proposed schedule for next steps

The report should never be just sent to the leadership team. The leadership team needs to be engaged in the feedback and adequate time allowed for discussion and questions.

What is critical is the action plan that comes from the assessment results. So, the process should be designed to make it clear that the assessment is not finished until the action plan has been completed. Many assessment processes stop at the feedback, but the best ones have a check-in place to make certain that an action plan has been agreed on and ownership assigned, typically within four weeks of the assessment. The quality of these plans is critical and they should be reviewed by the assessment team as part of the overall system.

In an ideal world, the leadership team is facilitated by one of the CI support team, through an action planning session that links to their strategic planning cycle. We do not want a CI plan and a separate strategic plan. The CI plan needs to be a key component of the overall business strategy and assessments should support this.

4.3.9.8 Lessons Learned Reviews and System Improvement

The system should also include a formal lesson-learned review for the assessment team and regular Voice of the Customer reviews. The focus should be:

What did we learn about how we planned and executed the assessment?
How can we improve the assessment process?
How can we add more value for the customer?
What do we need to change in the standard work?

Lessons learned reviews with the whole team and customers should also take place at least annually and ideally every six months in the first two years.

4.3.10 What Is the Communication Plan

We recommend involving the internal communications teams as soon as possible to broadcast loud and clear the purpose of the assessment system, how it will operate, and why it is important to the business. This cannot be done by email alone and gaining support at senior levels is important to success (see Chapter 1 on sponsorship). Some of the best examples we have seen include mini booklets that are easy to follow and that can be used by teams to understand the what, the why, and the how easily.

We have seen some teams try to keep the detail of assessment criteria a secret on the basis that people will try to "game" the result by solely focusing on the criteria. This is a mistake. It illustrates problems with the organization's culture in relation to trust and does not reflect the purpose

98 ■ *Why Bother?*

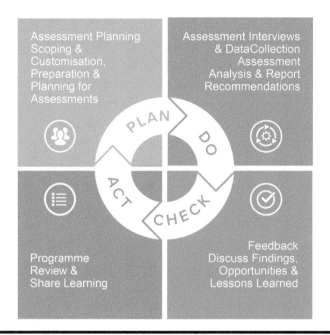

Figure 4.4 Plan, Do, Check, Act Cycle for Undertaking an Assessment.

of the assessment. There should be no big surprises and people need to understand the expectations at the various maturity levels.

4.3.11 Planning and Scoping an Assessment

The steps involved in planning each assessment should not be underestimated. We recommend a PDCA approach to each assessment, which is summarized in Figure 4.4:

Each assessment will have its own specific requirements. There may be a wide geographic spread of smaller teams, some language requirements, or even travel or visa restrictions. A practical approach is essential to guarantee that the assessment can take place with the optimum result for the customer.

4.3.11.1 Preparation and Planning of the Assessment

Some of the key points that need to be considered and built into standard work are:

Agree on purpose, process, and expectations with the site leadership team. Ensure appropriate advanced communications to the site workforce: not just an email.

Have a simple standard communication pack.

Agree on timing and an on-site support team—You will get pushback on the time and resources required but, as a benchmark, a Shingo Assessment is five people on site for three days.

Agree on assessment team roles—who will focus on which areas.

Ensure there is planned time for regular reviews and calibration—do not try to squeeze this into a five-minute check-in over a coffee break.

Agree on interviewees—they should be cross-functional and be representatives from all levels. Typically, around fifteen to thirty interviews. Ideally, get a list of names from various levels and assessor team nominations at random. For teams of up to two hundred people, then conduct fifteen to twenty interviews; for teams of more than two hundred people, then conduct twenty to thirty interviews.

Understand meeting schedules (especially visual management boards and huddles) and arrange to observe.

Agree on what data can be provided ahead of the interviews (e.g., financial results and any other KPI performance data such as customer feedback or engagement survey results).

Agree on a date for the leadership team feedback workshop and ensure it is locked in.

Schedule a date and time for the assessor team to conduct a lessons learned review.

4.3.11.2 Assessment Interviews and Data Collection

Some key things to ensure are in place before arriving on site are:

Have a clear and agreed interview schedule for each assessor.

Nominate a lead assessor to oversee the process and make final calls on areas of discussion.

Ensure detailed notes are made of each interview in a standard format.

Take time out to observe the workplace—for example, conduct a waste walk, observe huddles, or just stand and watch.

Collate the results into an assessment spreadsheet/tool.

Have a list of question prompts.

Use the pre-assessment work or previous assessment to determine areas of focus and to inform what areas to focus on.

Ensure the assessor team is consistent in the language being used and the standards being applied.

Always interview the most senior person first, or even in advance, and establish their views.

One powerful technique is to ask every interviewee an open question—we call this the magic wand question: If we could give you a magic wand and you could change anything at all to make the business even better, but you had to pick just one thing, what would you pick?

The answers to this question are often fascinating. They commonly reveal widespread underlying issues or opportunities and frequently some brilliant ideas.

4.3.11.3 Feedback and Action Planning

The assessment team should secure time in the leadership team's calendar for an interactive workshop to deliver their summarized findings and to give feedback as part of the assessment planning process.

The assessment team can support subsequent action planning with facilitation as required or, even better, coach the local support team to help them assist the leadership.

A useful way to frame expectations is to make it clear that the assessment is not considered complete until the action plan has been created by the leadership team.

4.3.11.4 Review and Best Practice Sharing

This is a very powerful Learning and Development opportunity. Our goal needs to be to constantly improve the quality and effectiveness of the assessment process. Some key things to include:

After each assessment, the team should review:
what went well
what could be better
what can be improved next time.

A couple of times a year all assessors should do a shared lessons learned review.

A process needs to be in place for customer feedback and improvement suggestions.

4.4 Key Takeaways

1. Always focus on who the customer is and what value they get out of the assessment.
2. The assessment system will never be perfect. It needs to evolve, adapt, and improve as lessons are learned and business maturity around a CI culture increases.
3. Building your own assessment system will increase understanding and promote the development of leaders and assessors across the whole organization.

Notes

1. John Bessant and Sarah Caffyn, "High-Involvement Through Continuous Improvement," *International Journal of Technology Management* 41(1), (January 1997). DOI: 10.1504/IJTM.1997.001705.
2. Chris Argyris and Donald A. Schön, *Theory in Practice: Increasing Professional Effectiveness* (San Francisco: Jossey-Bass Publishers, 1974).
3. Michael E. McGill and John W. Slocum, Jr, *The Smarter Organization: How to Build a Business That Learns and Adapts to Marketplace Need* (NY: John Wiley & Sons, 1994).
4. Hines and Butterworth, *Essence of Excellence*, 36.
5. Shingo Institute www.shingo.org.
6. Cheryl M. Jekiel, *Lean Human Resources* (New York: CRC Press, 2011).

Chapter 5

Why Bother Defining Behaviors and KBIs?

Chapter Summary

Key Behavioral indicators are essential to developing and maturing ideal behaviors that enable any organization to embed a sustainable culture of CI. They can be used across the organization in any area or function. This in-depth case study by Professor Peter Hines and STMicroelectronics explains why KBIs are critical and illustrates their application in both manufacturing and HR function.

Since the start of the 1990s, most organizations have sought to adopt some form of continuous improvement. For the majority, Lean has been the approach they have implemented, although many have adopted Six Sigma, Operational Excellence, or Agile as their preferred methodology. Much has been written about the respective strengths of these approaches. However, fundamentally they all share a technical, tool-based approach. While there are clear differences, many of these are cosmetic, such as in the terminology used. What is different is that each has its roots typically in different functions: Lean in Engineering, Six Sigma in Quality, Operational Excellence in Operations, and Agile often in ICT.

This chapter is written by Peter Hines, Irene Teo, Camille Pied, Kailash Chandra Joshi, and Ashish Chawla.

DOI: 10.4324/9781003185390-6

104 ■ *Why Bother?*

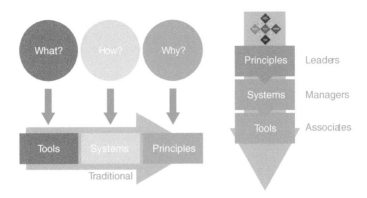

Figure 5.1 Tools, Systems, and Principles.

Not surprisingly, these approaches are generally gauged by way of resulting or lag Key Performance Indicators (KPIs) and technical assessments that often revolve around how mature the application of the tools are in their usage. For many, this is a good starting point for their improvement journey.

However, there are two major problems. First, true improvement, or what we might term Enterprise Excellence, requires a more systemic approach. Second, it requires a greater emphasis on people and culture. These two points are neatly summarized by Jon Alder in Figure 5.1.[1] As our knowledge of organizational improvement has evolved many organizations have progressed from a tools-based approach to a more systems-based approach, looking at systems such as Order Fulfilment or Strategy Deployment. However, at this stage, improvement is still largely viewed from a technical perspective. There are a small number of organizations that have evolved further and decided that they should adopt a more principled or cultural journey.

To do this, we need to look at improvement as involving more than just the technical tools and systems of what Toyota calls "Continuous Improvement" but also the people approach, or what Toyota calls "Respect for People." This is like the other side of a disconnected bridge, a side that many practitioners from the left-hand side of the bridge have only glimpsed in the distance. This having been said, many of the leadership and culture change approaches on the right-hand side of the bridge are well known to those from a human resource management or an organizational development functional background. Sadly, for them, the technical left-hand side of the bridge is equally distant. This means that often their inputs are not taken as seriously as they might be. What is needed is to bridge this gap (Figure 5.2).[2]

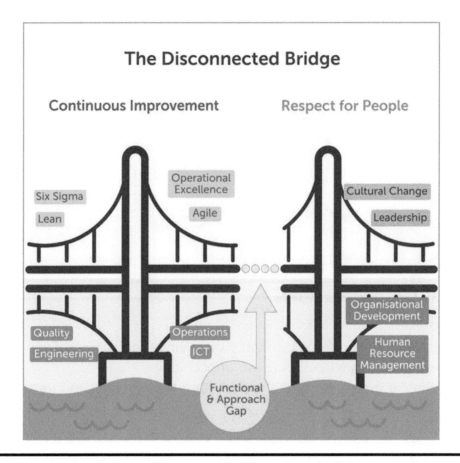

Figure 5.2 Enterprise Excellence and the Disconnected Bridge.

What is required is to create a business around a series of core operating principles that are designed to be both technically strong as well as culturally strong. Such a set of systems is described in Figure 5.3.³ Here, four systems are defined around a PDCA cycle:

Plan: Behavioral and Strategy Deployment
Do: Continuous Improvement
Check: Leader Standard Work
Act: Learning and Development

Such an approach, therefore, guarantees that the organization is:

- aligning both culturally and strategically
- improving by big step Discontinuous Improvement, small step Continuous Improvement, and end-to-end Process Improvement

106 ■ *Why Bother?*

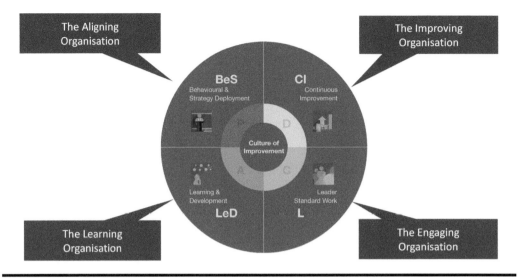

Figure 5.3 Core Operating Systems.

- engaging the workforce through recognizing their achievements and offering support where there are opportunities
- learning at every step of the journey through coaching and on-the-job learning

5.1 Culture and Behaviors

Much has been written about organizational culture and many theories have been presented. However, in our simple way of thinking, the culture of an organization is the sum of the behaviors of the people who work there. Change the behaviors and you change the culture. However, what is "behavior," as the term is often used very loosely? The English Oxford Living Dictionary[4] defines culture as "the way in which one acts or conducts oneself, especially towards others." Often, the term is used to describe any activity; however, we believe that it should be reserved for an "active" rather than a "passive" activity. In other words, things that can be observed, described, and recorded.

Consider the following activities:

- The team leader has a good idea of who is doing a good job in the team.

- The team leader knows how and when to recognize people within the team.
- The team leader recognizes the appropriate people in the team.

While the first two are good and necessary, only the last is a behavior using our rather strict criteria. Notice this is the only one that is active. This activity would clearly fit within the Leader Standard Work we discussed above. A further question: Is this an ideal behavior? Well, consider the following questions:

1. Is this behavior only relevant for one team leader or many?
2. How often does the recognition take place?
3. How timely is it?
4. How much effort and detail has gone into the recognition?
5. What form or forms does the recognition take?
6. Is this recognition well received by the recipients?
7. Does it lead to a change in their behavior or activity?

After this consideration, it might be possible to make this good behavior more ideal by developing it into:

All managers regularly recognize the appropriate people in their team in a timely way through a range of informal and formal impactful methods.

5.2 Back to Measurement

We have now described what behaviors are. The question now is: Can they be measured? Look at Figure 5.4. As you can see, less mature organizations tend to measure their progress through a series of "lag" KPIs such a profit and on-time delivery. As an organization's maturity progresses they start to introduce some "lead" KPIs, such as the maturity of improvement activity and the time people spend on improvement activity. The reason for this is that the achievement of these metrics is then likely to improve their lag KPIs. They will almost certainly stop measuring some of the lag KPIs that prove less useful to them to avoid KPI proliferation.

In the same way, very few organizations have started to measure Key Behavioral Indices or KBIs. Examples of this might include meeting

108 ■ *Why Bother?*

Figure 5.4 KPIs and KBIs.

Table 5.1 Lag KPIs, Lead KPIs, and KBIs

Target Area	Lag KPI	Lead KPI	KBI
What are you trying to achieve?	How might you measure that?	What is the key influence on making this improve?	What behavior is most likely to make this happen?
Improved Productivity	Units per person per day	The number of improvement ideas per person per month	Managers ensuring that in their Leader Standard Work that they spend regular time coaching people on idea generation and supporting implementation

attendance and adherence to standards. Here, they are measuring the types of behaviors that we discussed above or that are required for the organization to be successful. Again, the organization is also likely to reduce the number of less useful KPIs it measures to avoid measurement proliferation.

To illustrate this, we will look at a simple flow-through example (Table 5.1). Let us suppose that the organization in question was seeking to improve productivity. To measure this, it might use a measure such as the number of units produced per person per day. Then it might consider what is the most likely lever to improve this lag KPI. This might be the number of improvement ideas developed per person per month. This would therefore be the lead KPI. It might then consider what behavior (or behaviors) are most likely to improve this lead KPI. In this example, they have decided that it is managers ensuring that in their Leader Standard

Work they spend regular time coaching people on idea generation and supporting implementation. Hence, there may be two manager behaviors that are measured: the time they spend coaching every week on generating ideas and the time they spend coaching implementation. At this point, the behavioral measures are either yes/no or quantitative in terms of the time spent on the particular behavior. We will return to this measurement type at the end of the chapter.

What we have found by running this exercise many times is that the KBIs can at first instance look strange to the outside observer, but we have learned that a key part of doing this is that the people being measured are part of the development of the KBIs, so the measures are meaningful and impactful to them.

5.3 Developing a Whole System

Above we described Behavioral and Strategy Deployment as a core operating system of an advanced Enterprise Excellence organization. For brevity we will not review the Strategy Deployment part of this system (with the accompanying KPIs, for instance), but will concentrate on the Behavioral Deployment part and how this might be applied. This is for two reasons. The first is that Strategy Deployment is not new and is often done well. The second is to remain concise.

For some companies, behaviors are developed out of mission statements, business values, or visions. For others, they seem to appear out of nowhere, or perhaps a closed-door boardroom meeting. We believe that organizations should take a systematic approach to this as well as their subsequent deployment. There are several ways to do this. Figure 5.5 illustrates a simple eight-step approach, where the first four steps cover "behavioral formation" and the second four steps cover "behavioral deployment."[5] We will briefly describe this theory before reviewing an example of application at STMicroelectronics.

Step 1: The best starting point is a set of guiding principles or values such as those provided by the Shingo Institute.[6] Many, especially larger, organizations are required to work within a set of business values that have been laid down by the head office.

Step 2: As we mentioned earlier, many organizations are good at forming and deploying strategies. This is often logical and well understood

110 ■ *Why Bother?*

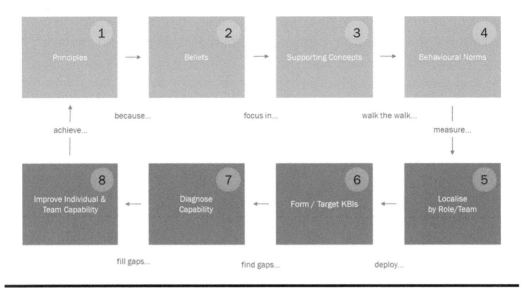

Figure 5.5 Behavioral Formation and Deployment.

by the workforce. However, in many cases, this does not lead to high levels of engagement. This is because they have deployed the HEAD (the logical strategy) and the HANDS (the deployment and tools) but have missed the HEART. There is no point deploying the logical WHAT if you do not create in people a belief that they want to do it. As Shigeo Shingo often said, "We have to grasp not only the 'Know-How' but also 'Know Why', if we want to master the Toyota Production System."[7] Hence, if we want to take the principles to heart, we should discuss what they mean to us and why we believe they are true.

Step 3: The Shingo Principles are quite high level, as are many company value statements. In some cases, this can mean that people cannot grasp them as easily as they would like to. However, with the Shingo Principles, there are also a set of "supporting concepts"[8] that provide greater clarity. For instance, what do the two "Cultural enablers" principles of "Respect every individual" and "Lead with humility" really mean?

WELL, this becomes clearer when the supporting concepts are provided:

Assure a safe environment—There is no greater measure of respect for the individual than creating a work environment that promotes both the health and safety of employees and protection of the environment and community.

Develop people—Through people development, the organization creates the "new scientists" that will drive future improvement. People development includes hands-on experiences where people understand new ideas in a way that creates personal insight and a shift in mindsets and behaviors.

Empower and involve everyone—For an organization to be competitive, the full potential of every single individual must be realized. People are the only organizational asset that has an infinite capacity to appreciate in value.

So, Step 3 is about creating a deeper understanding of what the principles really mean.

Step 4: Now that an organization has established what the principles really mean it can go about developing a set of behavioral norms so that they can "walk the walk." Here, they consolidate a set of guiding behaviors that they want to use in the organization.

This looks at first to be a straightforward exercise, but it is not easy! There are two difficult areas:

What is and what is not a behavior?
How do you make these behaviors as close to ideal as possible?

Step 5: The next step is to localize these by role or by team. We have seen many companies develop a set of behaviors. However, what is far rarer is the actual deployment of these behaviors in a meaningful way. Even when this is done, we find it is rarely done well.

Here, we are looking to deploy the behavioral norms to every person, every team, and, indeed, every recruit. There are three types of behavioral deployment that we might consider:

individual behavioral deployment
team-based behavioral deployment
behavioral-based recruitment

Over time it is useful to undertake each of these deployments. However, you might want to start with one—probably one of the first two. Later, you might also wish to focus on behavioral-based recruitment, as it is easier to work with people who already have the right behaviors than it is to change their behaviors.

Step 6: Here, we are seeking to measure the behaviors themselves. Are we regularly deploying these types of behaviors in a local environment? This can be done in numerous ways. The simplest is a plain yes or no within a given period, such as between weekly team meetings. However, more important than a plain yes or no (or green or red indicator) is the conversation that follows, as well as any appropriate recognition or corrective action.

Step 7: The seventh step is to diagnose the capability of the individual or team concerned. This might be achieved with a team, for instance, by measuring the number of yes or no answers (green or red indicators) over a period, such as a month or a quarter.

Step 8: From this collection of data and diagnosis it will be possible to then start working on the capability of the individual or team concerned through education, coaching, or observation of other individuals who might be exemplars.

In this chapter, we will look at a team-based deployment at STMicroelectronics and how they did this at their Singapore and New Delhi businesses. A wider discussion of individual behavioral deployment and behavioral-based recruitment can be found in Hines and Butterworth.[9]

5.3.1 Behavioral Deployment at STMicroelectronics

STMicroelectronics is one of the world's largest semiconductor companies with revenue in 2019 of US $9.56 billion, employing forty-six thousand people, and with eleven manufacturing sites. Approximately half of the manufacturing is in the Asia Pacific region. In this chapter, we will look at the behavioral deployment activity of the Ang Mo Kio manufacturing site in Singapore as well as the Noida, New Delhi Research and Development center in India.

5.3.1.1 Ang Mo Kio, Singapore

The Ang Mo Kio site has almost four thousand employees, meaning that any cultural change is likely to take time and considerable work. The Lean activity at the site, including behavioral deployment, started in 2016. It followed a review by the site General Manager who saw that, after many years of improvement on lag KPIs such cost, productivity, quality, and service, progress was reaching a plateau. Hence, a breakthrough approach across all levels of the organization was required if the site was to reach the next level and create a sustainable future. It was decided that this would be achieved through a Lean methodology including:

- a deep review and reboot of existing management practices, and
- the application of additional lean tools and methodologies.

This approach was captured in a new site vision:

Together, we provide our "customers" a "competitive" and "agile" manufacturing operation.
We strive to achieve "sustainable operational excellence" by leveraging on aligned behaviors, world-class processes, and innovation.

Hence, what was required was an emphasis on both sides of the Disconnected Bridge (Figure 5.2) to create an effective and sustainable approach. In terms of the management practices, a focus here was required to catalyze change and foster teamwork which would create a positive environment to capitalize on everyone's talent, and hence build and show the path to a successful future.

The behavioral work started in 2016 within a program called MAJU or Advanced in English (Figure 5.6). It can be described in three steps:

1. Our vision.
2. Our behaviors.
3. Our progress.

The first stage of this was creating a vision in terms of the most important management practices that needed to be addressed. This required three inputs. First, a planning day was held with the top eighty managers. From this, three hundred ideas were generated on practices, with a top ten developed. Second, input was gained from right across the site from a series of observations in manufacturing clean rooms, interviews, and a wider manager survey. This was brought together in a meeting with the site senior staff team and the MAJU steering committee.

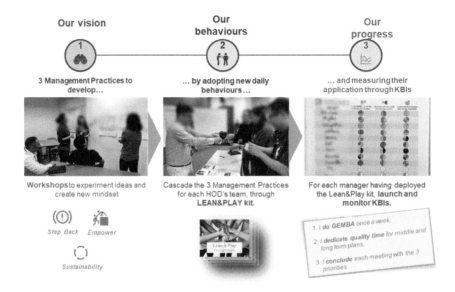

Figure 5.6 MAJU Management Practice Journey.

From this activity, three priority practices were developed. First, "Step Back" management, whereby people can react at an appropriate pace which needs to be rapid but not too rapid, as in traditional firefighting. This might involve, for instance, sufficient time to understand the root cause and develop appropriate countermeasures rather than applying temporary quick fixes. Second, "Sustainability," or the discipline to make sure that improvements are kept going, rather than switching from one approach to another without sufficiently embedding change. Third, "Empowerment" of everyone at all levels across the whole site (Figure 5.7).

In the "Step Back" it was then necessary to understand what these three management practices were, how they worked, and how good current practice was. To do this, brainstorming workshops were held with the eighty heads of departments. Then, each of these senior people was asked to take at least one action and practice with their team over the next one-week period. For example, one of the heads might work on empowerment by asking one of their team to chair a particular meeting. Actions were then developed. In this case, the manager would talk to their team member and discuss what support they might need for this to be successful. The success of these actions (yellow sticky note) or failure (pink sticky note) was then discussed at the next heads-of-department weekly meeting (Figure 5.8).

This experiment allowed the team to really understand the management practices and what they would entail. The MAJU coaches were then able to

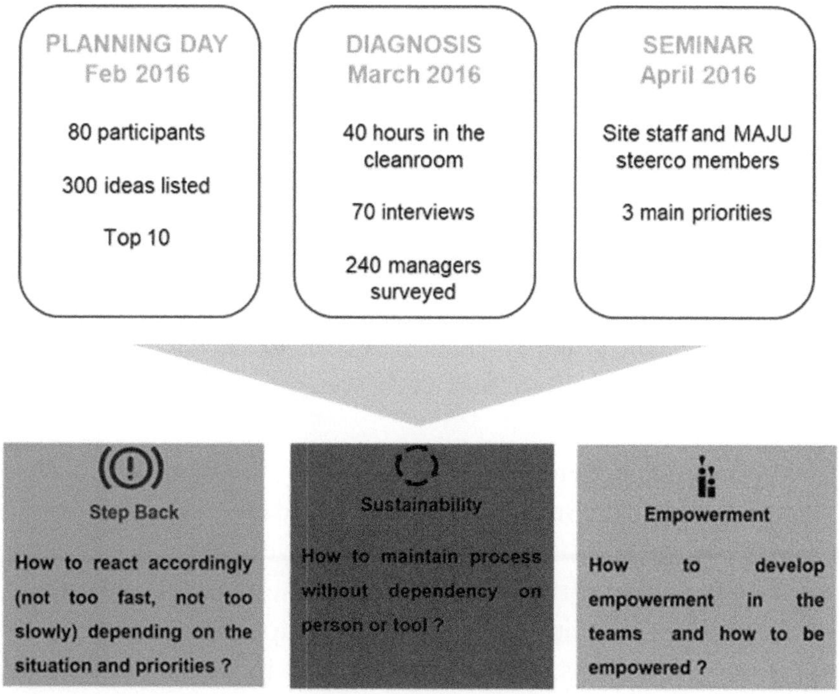

Why Bother Defining Behaviors and KBIs? ■ 115

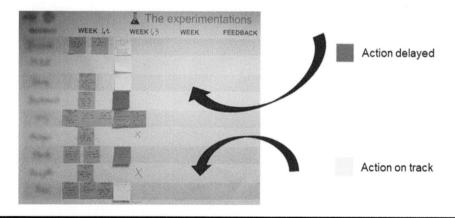

Figure 5.7 Developing the MAJU Management Practices.

Figure 5.8 Head of Department Experiment with the Management Practices.

develop a support-card kit set for each of the three management practices (Figure 5.9). This included a two-sided card of examples of what to do for different examples of empowerment.

Support kits were also developed for the Empowerment and Sustainability management practices (Figures 5.10 and 5.11). In the case

  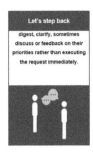

Figure 5.9 Support Kit Card Set for Step Back Management Practice.

of Empowerment, there were two aspects. First, as a manager, how do you empower people? Second, as an employee, how to be empowered by your manager and help your manager to empower you? This is because in Empowerment we need two hands to clap: the manager and the employee. With Sustainability, a set of key principles were developed for the sustainability of projects. This was not about increasing investment but more about the actions of people with what they could do to guarantee the sustainability of their projects.

The second stage was developing behaviors and cascading to the levels below. This involved the use of the three support kits and a video explaining what to do and what not to do, which were shown during already scheduled quarterly briefings. It also involved the use of a centrally located visual set of display boards called MAJU Corner (Figure 5.12), as well as a communication kit for heads of department called "Lean and Play."

Lean and Play was the support mechanism to help the eighty heads of department deploy the management practices to their own team. The key here was that this was not done by one of the MAJU team but by the manager themselves. This meant that the manager could explain the method; how they would, for instance, like to improve empowerment, their own personal commitment, as well as the commitment that the whole team needs to make. As the managers were to run this training themselves, it also greatly increased their own understanding. They were, however, supported by a MAJU coach or a member of the HR department.

The Lean and Play involved six sessions, with two on each of the three management practices. These were each ninety minutes and involved a series of practical exercises using posters and sticky notes—no PowerPoint

Figure 5.10 Support Kit for Empowerment Management Practice.

118 ■ *Why Bother?*

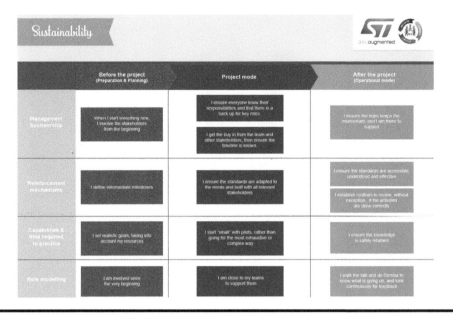

Figure 5.11 Support Kit for Sustainability Management Practice.

slides! They were run every two weeks, allowing time in between for practice. In a similar way to that described above, the team members made a commitment to action, practiced on the respective management practice, and shared the results of this action at the next meeting. This is illustrated in Figure 5.13.

The third stage in 2018 was then to measure the progress of the behavior change. To do this it was necessary to develop some KBIs against the three management practices. This was because the three management practices were rather general and, therefore, if they were to be measured, they needed to be made explicit in terms of specific routines or behaviors. Eight such routines were identified and, during a voting process at the annual management head meeting, top three were identified (Figures 5.14 and 5.15).

After this meeting, more detailed KBI cards were developed for each of the routines (Figure 5.16). These gave information about what the activity was, how to do it, what the key priorities were, and some tips.

The deployment process is shown in Figure 5.17. After the initial six ninety-minute training sessions, a further two-hour training session was developed to explain to local teams what KBIs were, the cards, how the template worked, and what challenges they might face.

This approach has now been applied to the eighty local teams across the site using the template shown in Figure 5.18. These are then deployed

Why Bother Defining Behaviors and KBIs? ◾ 119

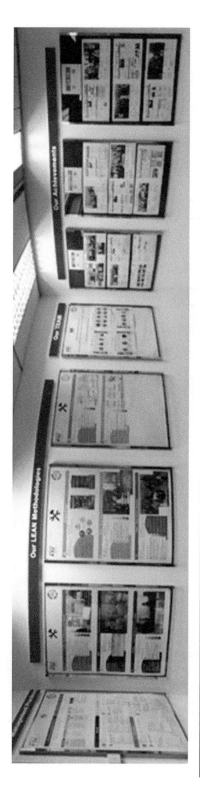

Figure 5.12 MAJU Corner.

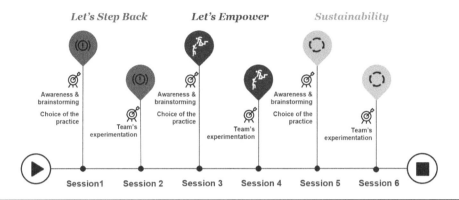

Figure 5.13 Lean and Play Process.

Figure 5.14 The Eight Routines (or Behaviors) that were Initially Defined.

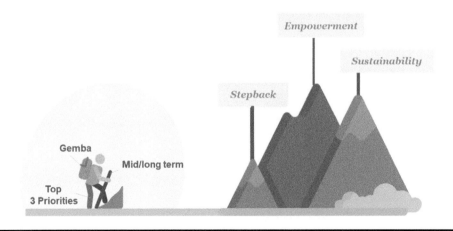

Figure 5.15 The Three Prioritize Routines and Three Management Practices.

Why Bother Defining Behaviors and KBIs? ▪ 121

Figure 5.16   KBI Cards.

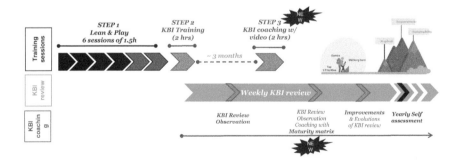

Figure 5.17   Lean & Play and KBI Process.

on their local visual management boards. A column is defined for each of the routines with summary behaviors indicated (How do I behave?). If this behavior has been demonstrated by the person concerned in a particular week, then one quartile of the circle in the respective column is

122 ■ *Why Bother?*

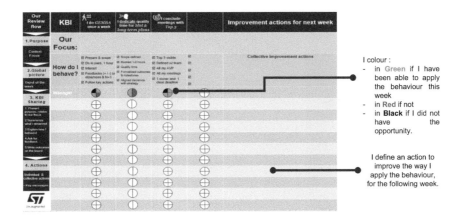

Figure 5.18 KBI Template.

colored green. If there is room for improvement then it is colored red, and if it was not relevant (e.g., due to holiday) it is colored black. These are then reviewed in weekly visual management board meetings, where the managers facilitate a discussion of why the measure is green or red. Where necessary, individual or collective improvement actions are recorded to be addressed over the following week. The chart is then filled in for a four-week period before it is wiped for the next four-week period.

After the first few weeks, the MAJU coach comes to observe team meetings and, if necessary, provides further coaching by way of a further two-hour session including the use of a training video. This is to help managers improve how they facilitate their KBI review. Just recently a KBI maturity assessment has been developed which is used to assess each team (Figure 5.19). This has three levels. Level 1 is that the team is "I practice," or, in other words, that the team is doing a weekly KBI review. Level 2 is "I practice well," which covers the maturity of their activity. This is designed to help the managers and teams to judge how well they are doing the KBI review. It is not designed to be used as a comparator between teams but more for the team's own benefit. Both Levels 1 and 2 are self-assessed, together with further observation and coaching by MAJU coaches.

Level 3 is, at the time of writing, under development. It will review the impact of the routines and associated behaviors. The idea here is to ask the recipients of the behaviors how they feel the routines are being carried out. This will, therefore, measure the impact of the routines. This is illustrated in Figures 5.19 and 5.20.

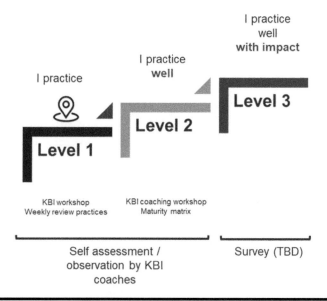

Figure 5.19   KBI Maturity Assessment.

#	Category	Scenario Selection				Score Max 4 pts
1	Format & Template	Template start with team manager	Disks updated	Color code respected (green - red - black)	Actions - due date - status updated	
		Selected KBI are in the FMT SG list	Kbi review duration approx 15'	Team KBI focus defined and visible	Attendance complete	
2	KBI review flow	Global picture (introduction and trend)	Review column by column	Presenters >1	Review actions & Recap outcomes of the review	
3	KBI sharing	Present purpose	Summarize what I observed	Explain how I behave	Explain the outcomes	
4	Contents & Outcomes	Discussions aligned with team KBI focus	Kbi standard card applied	Sharing of personnal experiences and discussing team improvements	Action plan defined and written	
5	Interactions	All audience attentive	Active participants > 50%; Spontaneous sharing, safe space	Feedback given on behaviour improvement	KBI review is a team discussion	
6	Manager's Facilitation skills	Manager impulses energy	Positive reenforcement	Focus on behaviour rather than technical topic	Manager encourages discussions around team improvement	

Figure 5.20   KBI Maturity Matrix.

5.3.1.2 Noida, New Delhi, India

There is no mandatory approach to behavioral and KBI deployment across STMicroelectronics. Each facility is free to follow its own path, although support is offered regionally. Hence, the approach at the Noida site in India, while similar to Singapore, has a number of differences. The people count at the site is 1330 and the primary activity is research and development, although there are also IT, sales and marketing, and support groups. The main challenge for Lean has been the application in an office research and development environment, with few available local benchmarks. Like other sites across STMicroelectronics, the Lean activity has its home within the HR function. At Noida, the HR team numbers fourteen people, and they facilitate the deployment through the cross-functional, thirty-two-member LEAN chaupal.[10] In this case, we will focus mainly on the use of behaviors and KBIs within the HR function itself. This is because the HR function was the pilot area, but, as in Singapore, the approach has been spread across the site.

The Lean journey at Noida started in 2014 when Lean was introduced by the Group Head of HR, Philippe Brun. At this point, they developed their first visual management board. Over the next five years, there was particularly good progress concerning the application of a Lean toolkit involving around half of the people at the site. The development and application of behaviors and KBI started in 2019, as it was important to develop a Lean culture through first identifying and embedding a culture of KBIs. This is reflected in their Lean Mission & Vision:

> To achieve Customer Satisfaction "above all else," through building a Flow efficient "Collaborative Culture," contributing to ST India transformation to becoming the **best** R&D center within ST.

The KBI approach has been quite fundamental in moving the site forward for leaders, especially in terms of their Leader Standard Work at the site. This was developed first in the HR function through a series of three full-day workshops, where a set of values were developed as well as a set of associated behaviors, and how these could be actioned and measured (Figure 5.21). The work was highly influenced by the work of Patrick Lencioni and his five dysfunctions of a team, as the team was the key focal point for the HR function. Making this point, Lencioni states, "Not finance. Not strategy. Not technology. It is teamwork that remains the ultimate competitive advantage, both because it is so powerful and so rare."[11]

Eight behaviors were identified, with four being prioritized. The result was four key HIVE values (a hive being a very strong home). These values stand for:

H—Help: Each person offers or seeks help from someone else.
I—Input: Each person shares directly with other team members something of importance, especially where the topic is difficult or sensitive hence avoiding issues being 'swept under the carpet' leading to festering discontentment.

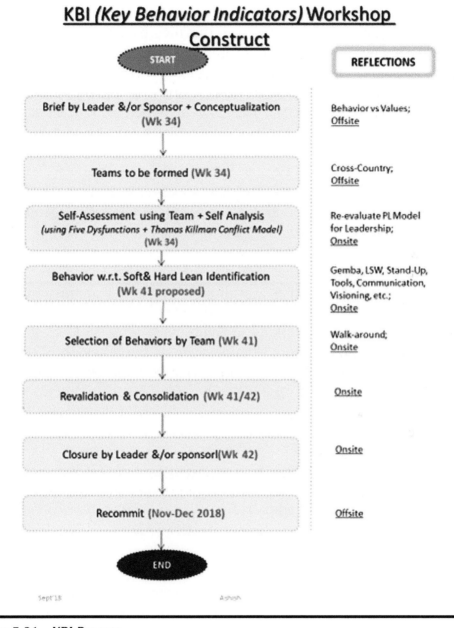

Figure 5.21 KBI Process.

V—**V**isual management: Each person has completed their own personal daily Visual Management Board.

E—achieving **E**xcellence: Each person shares an example of where they feel they have done something really well or excelled, hence leading to self-recognition and learning for other team members.

These KBIs were, as in Singapore, deployed into a visual management board (Figure 5.22). What differs here is that they have been able to use this as the central part of this team meeting. Hence, they have completely changed the focus of their weekly meeting in their HR obeya room to follow the flow described in Figure 5.1, with the first, and most important, part of a discussion of behaviors. This takes sixty percent of the total time in the meeting. Over the last two years, there has been a trend towards an increasing number of green markers on the board as behaviors have improved.

The next thirty percent of the meeting is spent reviewing more traditional improvements around processes such as talent management or organization development (Figure 5.23). Information is recorded in "to do" areas in each process, "in-process" work, as well as projects that have been "completed." Where completed, summary information is discussed about the "outcome/impact" as well as a "well done" given to team members. Details of each of the projects were recorded using "kaizen notes," seen at the bottom of the right-hand photo in Figure 5.23.

The last ten percent of the meeting is given over to a review of the (lag) KPIs in each of the HR process areas (Figure 5.24). Hence, the main focus of the meeting is on behaviors, with a good focus on processes and a quick look at KPIs, which should be going in the right direction as long as there is an appropriate set of behaviors and a focus on improving the main processes.

The whole team meeting activity is reviewed and improved on a quarterly basis and many aspects have been rolled out to other areas at the site, especially around the use of KBIs. This has also included the Lean chaupal itself. An example of this is from one of the IT teams, shown in Figure 5.25.

More recently, there has also been a site-wide review with the thirteen members of the HR team of the impact of KBI. It has shown very positive results. Each of the quantitative questions calls for a score of between one and four, so there is an almost uniformly positive view of the KBI activity, with each answer well over 3.0 (Table 5.2). There also seems to be good support to continue the activity with widespread introspection, although there are also some coaching opportunities in this area. The answers as to which KBI has been improved the most seems to vary considerably, although Visual Management (perhaps due to the time involved) and Sharing Directly (due to difficulty of having awkward conversations) seem to be the areas that were found to be the behaviors where people have been least successful.

Why Bother Defining Behaviors and KBIs? ■ 127

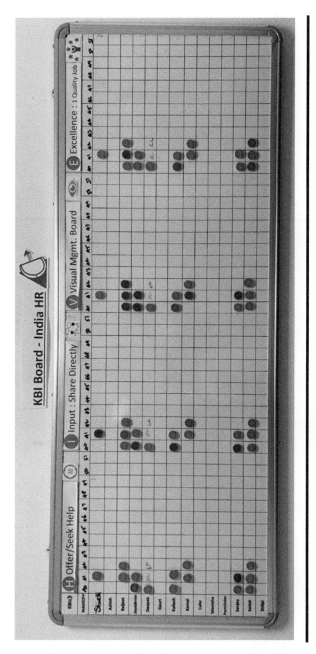

Figure 5.22 Visual Management of KBIs in HR.

128 ■ *Why Bother?*

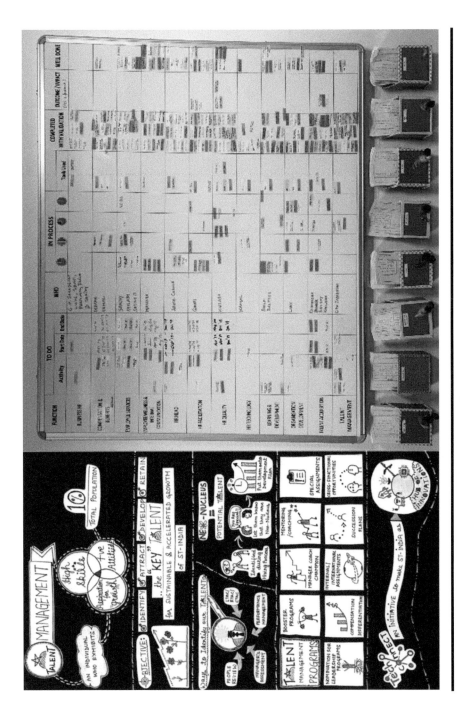

Figure 5.23 Visual Management of Process and Continuous Improvement in HR.

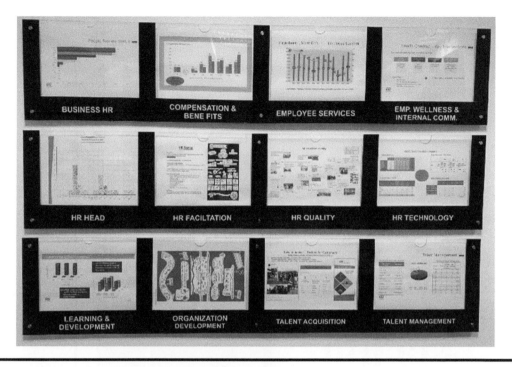

Figure 5.24 Visual Management of KPIs in HR.

5.4 Reflections and Conclusion

The application of KBIs has been reviewed in this chapter from both a theoretical and practical point of view. The examples from STMicroelectronics have provided a valuable insight into how this approach might be applied. What we have seen is a rapidly maturing approach in both the Singapore and Noida sites. These will almost certainly have moved on further when you are reading this chapter. What we can see is that this work has helped not only align the teams and the wider organization, but also focus improvement activity, engage the workforce, and, most of all, foster a learning culture.

One area that is currently under development is the type of KBIs. In both examples, the successful use of appropriate behaviors is recorded with a green mark, and behaviors with room for improvement being recorded as red. We might regard this as a Level 1 KBI; that is, a binary yes or no response. For some KBIs, such as "Gemba Walks" and "Offer and Seek Help," we might envisage recording this with the number of times it took place during a specific period. This might be regarded as a Level 2 KBI. However, what we have seen is that the real value here is less in

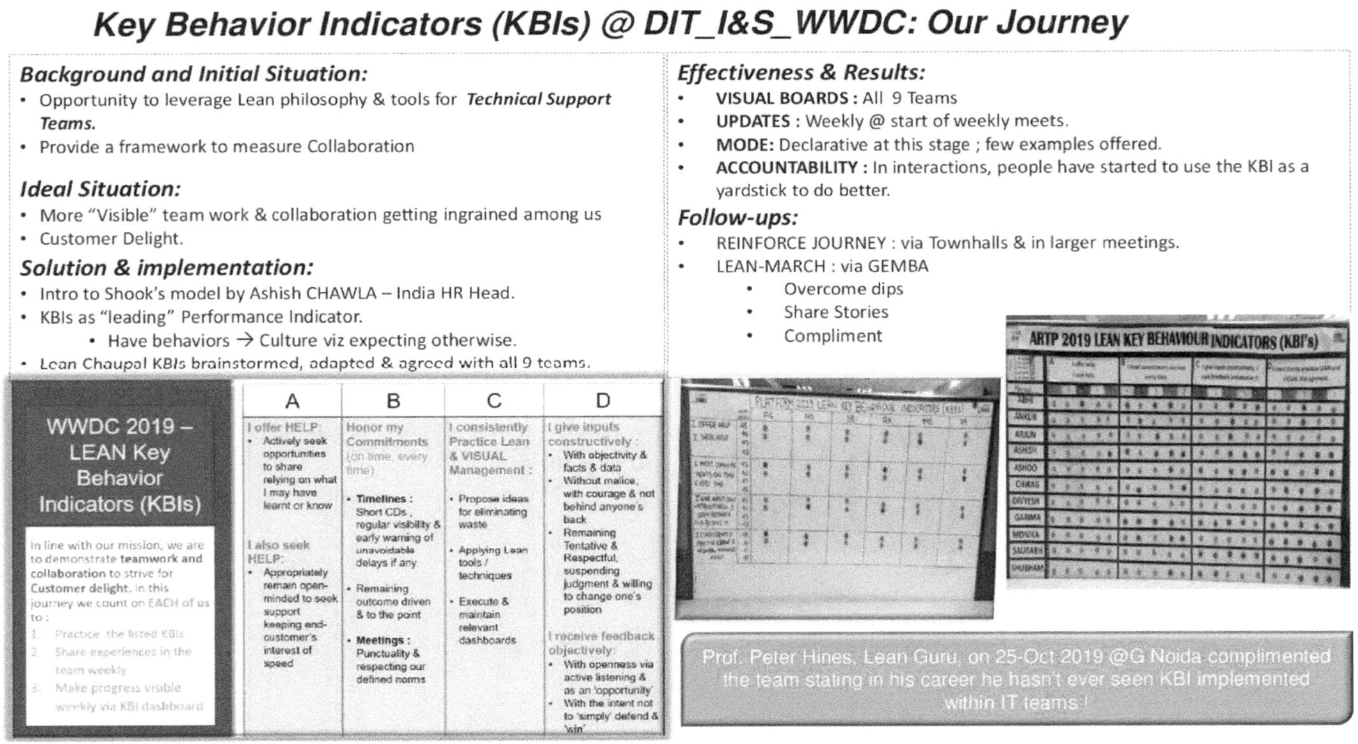

130 ■ *Why Bother?*

Key Behavior Indicators (KBIs) @ DIT_I&S_WWDC: Our Journey

Background and Initial Situation:
- Opportunity to leverage Lean philosophy & tools for *Technical Support Teams*.
- Provide a framework to measure Collaboration

Ideal Situation:
- More "Visible" team work & collaboration getting ingrained among us
- Customer Delight.

Solution & implementation:
- Intro to Shook's model by Ashish CHAWLA – India HR Head.
- KBIs as "leading" Performance Indicator.
  - Have behaviors → Culture viz expecting otherwise.
- Lean Chaupal KBIs brainstormed, adapted & agreed with all 9 teams.

WWDC 2019 – LEAN Key Behavior Indicators (KBIs)	A	B	C	D
In line with our mission, we are to demonstrate teamwork and collaboration to strive for Customer delight. In this journey, we count on EACH of us to: 1. Practice the listed KBIs 2. Share experiences in the team weekly 3. Make progress visible weekly via KBI dashboard	I offer HELP: • Actively seek opportunities to share relying on what I may have learnt or know I also seek HELP: • Appropriately remain open-minded to seek support keeping customer's interest of speed	Honor my Commitments (on time, every time) • Timelines: Short CDs, regular visibility & early warning of unavoidable delays if any • Remaining outcome driven & to the point • Meetings: Punctuality & respecting our defined norms	I consistently Practice Lean & VISUAL Management: • Propose ideas for eliminating waste • Applying Lean tools / techniques • Execute & maintain relevant dashboards	I give inputs constructively: • With objectivity & facts & data • Without malice with courage & not behind anyone's back • Remaining Tentative & Respectful, suspending judgment & willing to change one's position I receive feedback objectively: • With openness via active listening as an 'opportunity' • With the intent not to 'simply' defend & 'kill'

Effectiveness & Results:
- **VISUAL BOARDS:** All 9 Teams
- **UPDATES:** Weekly @ start of weekly meets.
- **MODE:** Declarative at this stage ; few examples offered.
- **ACCOUNTABILITY:** In interactions, people have started to use the KBI as a yardstick to do better.

Follow-ups:
- REINFORCE JOURNEY : via Townhalls & in larger meetings.
- LEAN-MARCH : via GEMBA
  - Overcome dips
  - Share Stories
  - Compliment

Prof. Peter Hines, Lean Guru, on 25-Oct 2019 @G Noida complimented the team stating in his career he hasn't ever seen KBI implemented within IT teams !

Figure 5.25 Example of KBI Deployment in IT.

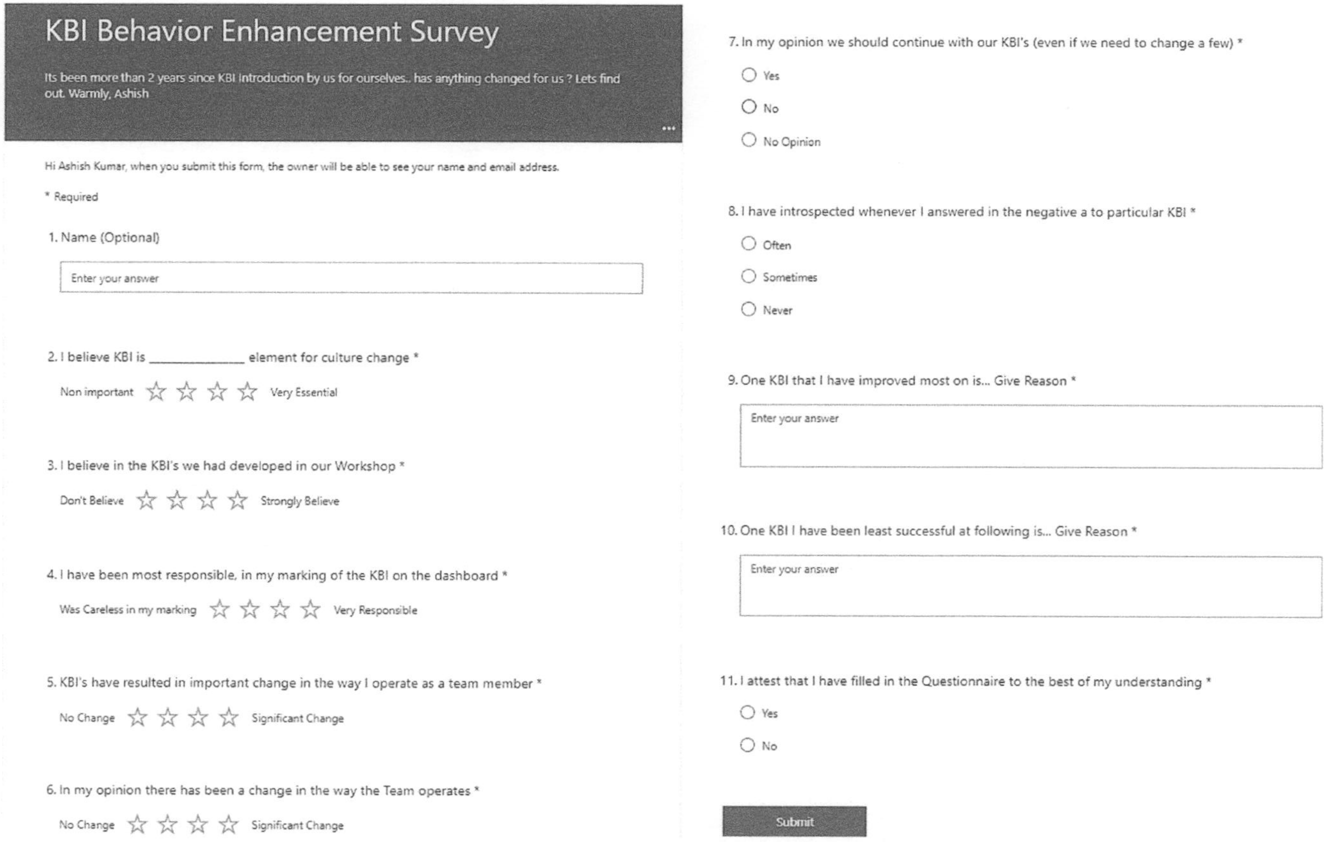

Figure 5.26 KBI Behavior Enhancement Survey, STMicroelectronics, Noida.

Table 5.2 KBI Survey Results in HR, STMicroelectronics, Noida

n=13	Score
I believe KBI is non important … very important element for culture change	3.92
I believe in the KBIs we had developed in our workshop	3.77
I have been most responsible in my marking of the KBI on the dashboard	3.38
KBIs have resulted in an important change in the way I operate as a team	3.23
In my opinion, there has been a change in the way the Team operates	3.38
In my opinion, we should continue with our KBIs (even if we need to change a few)	12 Yes 1 No Opinion 0 No
I have introspected whenever I answered in the negative a particular KBI	8 Often 5 Sometimes 0 Never
One KBI that I have improved most is … give reason (Illustrative answers)	Giving Feedback. was always not conscious of doing so earlier
	I feel this KBI platform has given me enough support to go and share inputs/feedback directly with a few of my team members. This is one area in which I feel I am uncomfortable otherwise.
	The one on excellence. I became conscious of my outcomes and felt I had to show something that could be valued as excellence. I would feel disappointed when for a few weeks in a row, I did not have something of excellence to share.
	Visual Management: Initially I used to rely on reminders to work on priorities, and sometimes reminders pop up when you cannot act on it immediately and I experienced it most of the time, the important priority getting missed. By Visual Management Board, it is continuous reminder or feedback to individual or team to act, and thereafter I started VSM with dates. No, I noticed 99% of the priority is closed before the due date.

Table 5.2 Cont.

One KBI I have been least successful at following is ... give Reason (Illustrative answers)	Visual Board. have been lazy
	Sharing feedback with others (I am open to take feedback). Though it has improved over a period of time, could be better. Reason. I lack the skill "how" to share feedback with HR colleagues, especially with senior colleagues. People rejected my feedback multiple times in the past.
	• Having a personal Visual Management board. • Have been making a 'Tasks to do' list every day for years and I feel they are effective, even following up on the smallest task. • Putting all tasks on board (particularly small ones) may not be feasible and therefore, may get lost/forgotten. • There could be notes, etc. pertaining to tasks that I can have in the same single notebook. It's easy to carry to other meetings and home if required.
	Visual Management is something I have not been able to follow—especially during the WFH scenario.
	Help—seek/offer—can do more and seek more help from other colleagues
	Visual Board. Maybe being a single person in my function I have not been able to successfully do it.

the simple recording but more in the conversations that might result. This leads to either recognition of the person responsible or an opportunity for improvement for the individual and their behavior.

Taking this idea a little further, we might envisage Level 3 KBIs that are based more on the **quality** of the behavior rather than whether it has been applied or the number of times it has been applied. Consider the example in Figure 5.27. Here, we take a theoretical example of a behavior norm, such as "we look around us for new ideas and ways of doing things." This is then deployed to a local team, who might interpret this as "regularly proactively spend time to look for internal and external ideas/practices." What is important here is how well this behavior is employed, and hence it might be possible to develop a set of questions that might be used by

134 ■ *Why Bother?*

Behaviour	Team Member behavioural deployment	What should we see/hear from Team Members?	What Coaching should we see from the team Leader?
We look around us for new ideas and ways of doing things	Regularly proactively spend time to look for internal and external ideas/practices	Team Member is dedicate a regularly amount of their time to learn from other members and internal teams Team Members making external benchmarking, customer visits and other learning opportunities	How often do you come up with a new idea? When was your last new idea? Where do you get inspiration for new ideas? Who is really good at this and what could you learn from them?

Figure 5.27 Level 3 KBIs Framework.

the team leader or their line manager as part of their Leader Standard Work routine. We might also be able to identify the coaching that should be provided to the team member by the team leader. This, of course, could be observed by the team leader's line manager so that coaching of the coacher might also be possible, again as part of a Leader Standard Work regime.

Another reflection is that the review and development of behavioral deployment can be likened to the training evaluation approach developed by Don Kirkpatrick and further enhanced by James and Wendy Kirkpatrick,[12] as shown in Figure 5.28. This involves four levels: reaction, learning, behavior, and results. So, for instance, in one of the KBI review sessions we describe above:

- The "reaction" might be scoring a particular behavior as green or red.
- The "learning" might be the specific reflection for the individual in what they have learned by practicing the behavior as well as the learning for the team when it is discussed at the team meeting.
- The "behavior" might be any specific changes in the quantity or quality of the behavior as a result of the review and subsequent coaching.
- The "results" might be the tangible results owing to a change in behavior; for instance, the lag KPIs in the Noida case.

Following this learning analogy, it might also be interesting to consider the further two levels to the Kirkpatrick model, as suggested by Phillips and Phillips;[13] namely, the Return on Investment of the work and the other Intangible Benefits. In this rapidly developing area, perhaps that is one for the future.

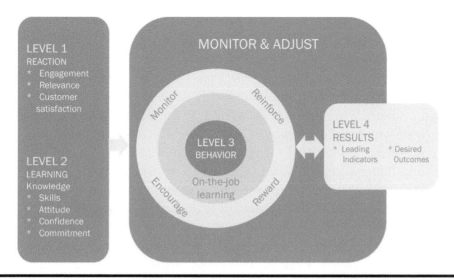

Figure 5.28 The New World Kirkpatrick Model.

5.5 Key Takeaways

1. Almost all technical (Lean, Six Sigma, and Agile) improvement programs that we have seen do not adequately address the human element of creating a sustainable transformation culture change, as the people elements are underestimated and often missing because they are usually driven by HR and are outside the remit of the program.
2. It is possible to combine these behavioral aspects in improvement programs, but it is not easy and takes considerable effort as well as a human-centered mindset.
3. The two case studies provided in this chapter illustrate approaches that can be applied to this integration, which have been applied across two large sites with considerable input from the HR function within STMicroelectronics.

Notes

1 Adapted from Jon Alder, Former Director, Group Lean Enterprise, Rexam.
2 Peter Hines and Cheryl Jekiel, "An Introduction to the People Value Stream," *People Value Stream*. Last accessed March 1, 2021. www.PeopleValueStream.com.
3 Hines and Butterworth, *Essence of Excellence*, 191.

4. *English Oxford Living Dictionary.* Last accessed 2017. https://en.oxforddictionaries.com/definition/behavior.
5. Hines and Butterworth, *Essence of Excellence*, 46.
6. Shingo Institute, *Shingo Guiding Principles.* Last accessed October 28, 2020. https://shingo.org/shingo-model/.
7. Shingo Institute, Why "Know your Why." Last accessed March 1, 2021. https://shingo.org/why-know-your-why/.
8. Shingo Institute. *Shingo Model.* Logan: Shingo Institute, 2017.
9. Hines and Butterworth, *Essence of Excellence*, 36–84.
10. A chaupal is a traditional committee of village elders in rural Indian villages.
11. Lencioni, Patrick, *The Five Dysfunctions of a Team: A Leadership Fable* (New York: John Wiley & Sons, 2010), 9.
12. James Kirkpatrick and Wendy Kirkpatrick, *Kirkpatrick's Four Levels of Training Evaluation* (Alexandria, Virginia: ATD Press, 2016), 11.
13. "Eleven Reasons Why Training and Development Fails … and what you can do about it." *Training*. (2002). Last accessed November 12, 2020. https://roiinstitute.net/wp-content/uploads/2017/02/Eleven-Reasons-Why-Training-and-Development-Fails.pdf.

References

English Oxford Living Dictionary. Last accessed 2017. https://en.oxforddictionaries.com/definition/behavior

Hines, Peter, and Cheryl Jekiel. "An Introduction to the People Value Stream." Last accessed 2020. www.PeopleValueStream.com

Hines, Peter, and Chris Butterworth. *The Essence of Excellence: Creating a Culture of Continuous Improvement.* Caerphilly: S A Partners, 2019.

Kirkpatrick, James, and Wendy Kirkpatrick. *Kirkpatrick's Four Levels of Training Evaluation*. Alexandria, Virginia: ATD Press, 2016.

Lencioni, Patrick. *The Five Dysfunctions of a Team: A Leadership Fable*. New York: John Wiley & Sons, 2010, p. 9.

Phillips, Jack, and Patti Phillips. "Eleven Reasons Why Training and Development Fails … and What You Can Do About It." *Training*. (2002). Last accessed November 12, 2020. https://roiinstitute.net/wp-content/uploads/2017/02/Eleven-Reasons-Why-Training-and-Development-Fails.pdf

Shingo Institute. *Shingo Guiding Principles* (2020). Last accessed October 28, 2020. https://shingo.org/shingo-model/

Shingo Institute. Why "Know your Why." Last accessed March 1, 2021. https://shingo.org/why-know-your-why/

Shingo Institute. *Shingo Model*. Logan: Shingo Institute, 2017.

Chapter 6

Why Bother Focusing on the Type of Conversations People Have?

Chapter Summary

It is clear from Kevin's chapter that how we have conversations is far more powerful than the words used. Indeed, the same words can be used to "tell, suggest, or ask," depending entirely on what "voice" we use. It also clearly shows how the types of conversation and voices used change significantly as a CI culture matures. As such, any behavioral assessment system needs to be able to assess how people are using their voices and if they are applying the appropriate dialogue in a skillful way.

While there is a lot of focus on behavior, one critical aspect that is often overlooked is the importance of the type of conversations people have. Some people are naturally gifted communicators but many of us are not. We need to work hard and practice effective communication skills. We also need to understand that various kinds of conversation are needed in different contexts and think carefully about how we should use our "different voices." The in-depth chapter below from Kevin Eyre details some excellent insight into this.

6.1 Sustaining Enterprise Excellence: Shift the Dialogue, Shift the Culture

6.1.1 The Importance of "Talk"

This chapter is a contribution to the debate on sustainable cultures of Continuous Improvement. It asserts that the use of language, of dialogue, of "talk," a free resource to leaders of organizations, is an overlooked and under-leveraged way of addressing a widely acknowledged problem, namely, that Business Improvement initiatives often fail to sustain. Blame for this failure is often laid at the door of "behavior" (since "the tools" always work) and especially, leadership behavior. And yet, "behavior" is a huge catch-all term often be-devilled by poor or abstract definition. Rather than attributing this failure (or indeed success) to sustain to the role of behavior, perhaps it would be better to look to the role of 'talk?'

What we say and how we say it has a profound, immediate, long-lasting, and continuous impact on those around us, maintaining and changing, by a small degree the quality of relationships, our perceptions of self, and the cultures in which we work[1].

Scholarship on language as a field of study is scattered across various disciplines; linguistics (including conversation analysis), sociology, philosophy, and anthropology doing the best job at housing the enormous range of astonishing research and intelligence in this field. Psychology also has a say but for the field that has most influenced thinking about "management and organization" for the past 100 years, psychology has had more to say about other important themes. In recent years, most notably, it has brought into focus "positive psychology" and "emotional intelligence" in which, although the present, language has had a minor role in the way these drama's present.

SoundWave seeks to play its part in redressing this balance. Our analysis of what people do with one another begins with the way they talk, with what and how they say things, and of the effect that these different ways of talking impacts culture and performance. We enjoy the face-to-face experience of seeing your gestures; we want to *hear* your thoughts; we need to be sensitive to and work with your feelings and the best way we can do this is through the miracle of the most human capability of all, our talk.

6.1.2 Continuous Improvement and Talk

Developing cultures of continuous improvement has long been and remains an aspiration of leaders and managers in today's organizations. Evolving this culture requires considerable human intelligence and is, at the risk of trivialization, 'a journey'; overnight transformation is indeed a fantasy. The attempt at creating a culture of this type is one domain (of many) in which dialogue holds a key to determining the success of the endeavor. We will assert that particular types of talks have either a positive or a negative effect on the success of the CI journey and that the type of talk necessary is intuitively understood by managers and is furthermore, learnable by them.

6.1.3 Modelling the CI Journey

In determining "the CI journey," there is bountiful research to draw on. The published work of Stephen J. Spear (*Learning to Lead at Toyota* 2004) sets out a compelling argument, that is, highly evolved cultures of continuous improvement, managers operate as coaches, as "scientists," and as teachers, rather than as domineering fire-fighters. The work of Mike Rother (*Toyota Kata* 2010), amongst others, elaborates and structures this point of view further. In its most simplistic form, in cultures of CI, the enterprise runs effectively, managers relate with consideration, and performance outputs are radically improved.

Conceptually stated, the journey to CI (Phase 4) is depicted in Figure 6.1.[2] The journey to a sustainable CI culture.

1. A pre-improvement phase (highly variable performance and process instability).
2. A phase of change and initiation (less variable performance and increasing process stability).

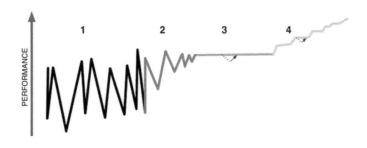

Figure 6.1   The Journey to a Sustainable CI Culture.

3. A phase of conformance or 'standardization' (of consistent performance and process stability).
4. A condition of continuous improvement (of improvement from the base of stability).

Simply put, accompanying each phase is a set of changes involving purpose, process, tools, and people. (These are set out more specifically in succeeding sections of this chapter)

6.1.4 Managers' Understanding and Experience of this Journey

Managers, it seems, have a good intuitive and experiential understanding of how these phases differ one from another. The table illustrates this very clearly. Over the past five years, we have asked over five hundred managers and leaders in workshop settings to answer a series of questions that focus, in particular, on the "people side" of the improvement journey.

Our questions have deliberately asked not just about "behavior" but about "talk." Not just what do people "see" and "do" but what do they "hear" at an environmental level and what "voices" do they discern amongst individuals in conversation with each other.

In the same way that clear distinctions can be identified by phase on the other questions, what people hear, is also clearly differentiated; it seems that people talk differently depending on the phase of CI evolution that they are in. This is summarized in Table 6.1.

Our findings of course make perfect sense. It is not possible to deliver any type of behavior without its corresponding form of talk. The model in Table 6.2, which takes a commonly desired "behavior" (often defined at Phase 2 as part of the transformational effort) specifies just how much of the delivery of this "behavior" can be done *with* talk and *without* talk.

When we view the matter in this way it seems like common sense. In so many situations not only is there a balance between talk and action but many times, talk *is* the action!

6.2 Talking, Phase by Phase

The rest of this chapter seeks to establish a relationship between the four broad phases on the journey to a culture of enterprise excellence and the type of talk that might typify those phases. To explain our argument, we

Table 6.1 Summary of Observed Behaviors at Different Maturity Levels

	Phase 1 into 2 — High Variation	Phase 3 — Stability	Phase 4 — CI
What does the environment look like?	Uncontrolled; Chaotic; Problems repeat; Closed; Fearful; Lagging KPIs; Ad hoc	Structured process and procedure; routines; Explicit goals and objectives; Leading KPIs; Structured Problem Solving; clear R&R	Low defects; energetic; inclusive; Stable and changing; Structured Problem Solving; Experimental
What does the environment sound like? What is 'the sound of the site?'	Complaints; Negative; Panicked; Vocal; Stressful; Frustrating; Emotional; Confused; Excuses; Noisy	Calm; Conversational; Data-driven; Drumbeat; Connected	Positive; Ideas; enthusiastic; Open; Humming; Buzzing; Trust; Common language
What is the role and behavior of managers and leaders?	Firefighting; Reacting; Not delegating; Command and control; Instructional; Unclear and uncertain	Accessible; Forward-looking; In-control; ensuring accountability; Holding standards	Strategic; Innovation; Challenge the status quo; KPIs; Coach; Inquiring and observing processes
What is the role and behavior of employees?	Headless chickens'; Absent; dis-engaged; Firefighting; 'Child-like'; Siloed; Reactive; Passive	Following system; role clarity; Collaborative; Appreciative; holding standard; Cross-functional; Providing feedback	Ownership; Experimentation; Accountable; Self-managing; open to change; empowered; Driving CI; Participating
If you coach, what do you coach for?	Task completion; under-performance; clarity on what the standard should be	Understanding; Personal Development; Confidence; Awareness; Improvement ideas	Improvement; Best-in-class; Evolution; Success; Deep understanding; helping others to become coaches
What 'voices' predominate?	Telling; Criticizing; Attacking; Blaming; Punishing; Shouting; Preaching; Advocating	Suggesting; Explaining; Advising; Diagnosing; Challenging; Critiquing	Asking; Inquiring; Challenging; Probing; Diagnosing; Explaining
Managers lived experience of each phase	High — many have spent time here	Medium — most have spent some time here	Low — some have spent some time here

142 ■ *Why Bother?*

Table 6.2 Summary of Behavior Needing Conversation

TRUST	100% Action	50/50	100% talk
Open and honest two-way communication			X
We honor our commitments and maintain agreed standards of work and behavior		X	
Issues will be flagged and addressed early		X	
We remain a respectful and open work environment, where questions can be raised openly, listened to, and answered in a reasonable time and manner			X

Figure 6.2 SoundWave Model.

have used the SoundWave model (shown in Figure 6.2) which identifies three clusters of three verbal strategies (or nine "voices") which people tend to use across their verbal interactions. More details are given at the end of this chapter in Appendix 6.1.

6.2.1 Pre-Improvement: Phase 1—The World of High Variation

From the vantage point of the system we are about to describe, Continuous Improvement seems a long way off, but this "world of high variation" as illustrated in Figure 6.3 is the starting point for many organizations.

Why Bother Focusing on the Type of Conversations People Have? ■ 143

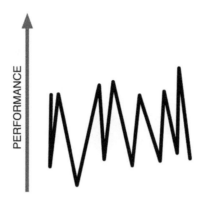

Figure 6.3   High Variation Phase.

The complexity of randomly developed processes causes high degrees of variation in service provision, product design, and service creation and delivery.

Many problems exist, often with many root causes most of which are not well understood. Instead, the complexity and associated variation have given rise to a culture commonly referred to as 'fire-fighting.'

From within this crucible, strong leaders emerge who, by dint of personality, address the problems and keep the wheels turning, but often at great cost. These cultures can be highly political if not toxic.

The randomness of a process produces randomness in behavior which, in turn, re-enforces the randomness of the process. It is a requirement that fires need 'fire-fighters,' but it is equally true that fire-fighters need fires!

Characteristically, it is like this …
Our pre-CI days were characterized by powerful regional barons who did what they could to protect their patch. Loyalty got rewarded; its opposite punished. There was no sign of anything approaching CI.
Director—Automotive Sector

Typically, the "sound of the site" might be captured in a SoundWave profile as shown in Figure 6.4:
In this type of culture, the use of language is strong, loud, and easily discernible. There is much positioning for power. There is much seeking of

144 ■ *Why Bother?*

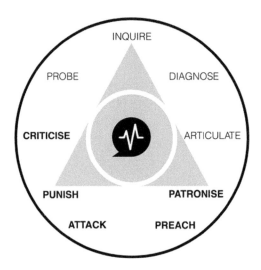

Figure 6.4 Typical SoundWave Profile at High Variation Phase.

control. There is an over-reliance on the strength and force of expression. There is a relative under-reliance on the voices of asking or the presence of active listening.

The noise of a pre-CI environment reflects the system in which it is bound. Occupying the bottom portion of the SoundWave model there is a tendency not just to use the cluster of Suggesting and Telling verbal strategies, but to accentuate them.

There is no coaching in this environment. Interaction is muscular.

6.3 Developing the Required Capability

At a macro level, significant and clear changes need to be made to the style and content of interaction between people if the culture is to change. It will not be enough to identify new values and define new behaviors although this will be necessary; such changes need to be lived and while my gestures, availability, presence, and 'positive attitude' will contribute to realize these changes, none of them will have quite the impact necessary without the skillful and authentic dialogue that accompanies them.

In essence, the range of my verbal style needs to broaden; the emphasis of my verbal style needs to sharpen and the capability to coach needs to deepen. The 'continuum of conversation' model shown in Figure 6.5 sets this out.

Why Bother Focusing on the Type of Conversations People Have? ■ 145

Style	TELL			SUGGEST				ASK		
Problem situation	Organization Crises	Standards Non-adherence	Stagnation	Disagreement	Skill or capability limitations	Conflict	Complexity	Participation	Career development	Ownership	Personal crises
Role	Commander	Instructor	Agent Provocateur	Negotiator	Trainer	Mediator	Consultant	Facilitator	Mentor	Coach	Counsel
Outcome	Momentum	Compliance	Agility	Compromise	Capability Development	Resolution	Simplification	Engagement	Growth	Accountability	Recovery

Figure 6.5 The Continuum of Conversation.

146 ■ Why Bother?

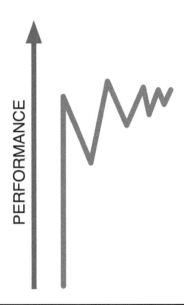

Figure 6.6   Phase 2 Reducing Variation.

Whereas a Phase 1 culture will hear talk operating at the Telling end of the spectrum (as indicated earlier), a Phase 4 culture will hear people interacting up and down the range of the spectrum as the situation demands it. We refer to this as "verbal fluency" and it becomes a focal point for developing improved interactional capability.

6.3.1 Change and Initiation: Phase 2—Reducing Variation

The problem of variation (and its associated culture) has tipped over into deep dissatisfaction with performance. There is recognition of a need for change and the search for solutions has led to a consideration of methods that have an impact across the whole organization. This is illustrated in Figure 6.6.

> This is new territory for the organization and the risks during these early attempts at change are high. Many organizations report 'failed initiatives.' However, at some point, the organization finds a way of making improvements stick and perhaps stumbles on the realization that two related problems exist—variation as the scourge of improvement and leaders as either helpers or hinderers.

Why Bother Focusing on the Type of Conversations People Have? ■ 147

The target for learning is clear and often centers on the application of tools. Their importance is huge at this stage. A bundle of new methods (and corresponding jargon—Kaizen Events, Value Stream Mapping, and so on) appears. Projects are initiated. New roles are identified and filled. Power shifts towards the front line and managers begin to understand that just 'demanding performance' is no longer enough. Now they need to create a system in which people can perform and adopt a form of dialogue and a mode of interaction that is considerably more facilitative.

Characteristically, it is like this …
As we tried to develop our own CI culture and capability, we had struggled in three main areas—sustainability; developing CI principles and processes as the day job rather than as a project; our lack of understanding and leadership in dealing with rapid change. It took several attempts and a combination of success and failure before we came close to making it work.

Director—Retail Operations

Typically, the "sound of the site" might be captured in a SoundWave profile as shown in Figure 6.7.

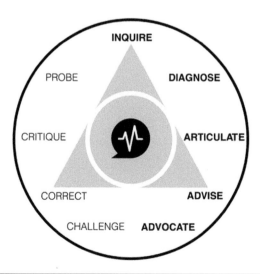

Figure 6.7   SoundWave Profile at Phase 2.

Dialogue has progressively shifted from Phase 1 to more constructive verbal strategies clustered around more socially safe voices but largely de-emphasizing those concerned with Control. Dialogue is heard to be persuasive, positive, and upbeat in a call for practical ways of making change.

The injection of new expertise (often consultants) as a mechanism for change combined with change vehicles which very often take the form of projects and project teams, tends strongly towards the use of the "suggesting" voices. In particular, the verbal strategies of "advocacy" and "advice" serve to encourage and influence the adoption of new methods. But even here there is room for the mentoring and coaching of CI environments, a chance to begin the drive towards a culture of ownership.

6.4 Developing "Phase 2" Capability

The capability required to make Phase 2 work carries risks and challenges. In Table 6.3, we outline a number of these and from our experience, propose ways of addressing them.

Table 6.3 Risks, Challenges, and Counter Measures (Phase 2)

Risks and Challenges	Countermeasures
Holding on to the directive leadership style characteristic of Phase 1	Raise awareness of the limitations of old styles through close 1-2-1 coaching and feedback. Plan 'experiments' using required styles. Assert new behavioral demands.
Not understanding clearly enough the end goal of a culture of enterprise excellence (Phase 4)	Provide education and exposure to progressive CI environments.
Not understanding the purpose of Phase 2	Build an understanding of this development theory.
Being unaware of the level and type of interactional skills needed	Build sensitivity to context, a conscious understanding of personal tendencies of interaction and skills via coaching, training, and real-time feedback. Increasing conscious awareness of the natural use of talk using appropriate tools such as the SoundWave self-perception assessment.

6.5 The Specifics of Talk

This is also the phase at which the organization re-sets its ambitions and becomes alert to the relationship between the performance outcomes it desires, the practical changes it needs to make to the enterprise and to the "values and behaviors" that will govern its changes. A new language emerges whose purpose is to provide guidance for the required and desired change in behavior—respect, humility, and collaboration for example. For organizations who take this matter seriously much work is needed to create meaning from this language and to avoid the possibility of cynical responses.

For organizations that take this path (recognizing as they do that the tools by themselves won't work unless the culture is capable of accommodating them), there is often a hard but ultimately rewarding road ahead.

6.5.1 Conformance: Phase 3—Ensuring Process Control

The benefits that accrue from radical reductions in variation promote a desire for standardization. (Standard work, 5S, management systems, Visual Management amongst other tools are characteristic of this stage) The particular challenge is in holding and correcting these standards and this is, at first, often difficult. Processes don't stay the same, they either get better or they get worse and so they need attention, especially when standards slip. This is visualized in Figure 6.8.

> At its worst, this is an extreme world where 'shadow boards for staplers' belies a misunderstanding of purpose and destroys efforts at realizing human creativity. At its best, it represents a 'call to arms' to engage human minds in a structured attempt at problem resolution laying the foundations for innovation and change.

Characteristically, it is like this ...
We had struggled with achieving consistent and reliable ways of working for some considerable time. Getting the process itself well-defined took us some of the way but it was insufficient. Natural, early failures to get things right almost always resulted in heavy handed

150 ■ *Why Bother?*

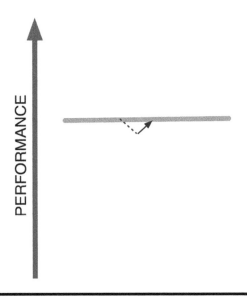

Figure 6.8 Conformance and Standards.

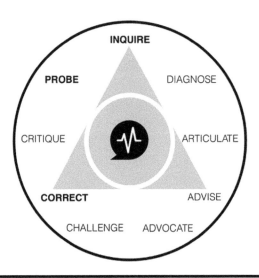

Figure 6.9 SoundWave Profile at Phase 3.

management. We had to learn to understand and help to correct and to not punish or walk away.

Director—Automotive Sector

Typically, the "sound of the site" might be captured in a SoundWave profile shown in Figure 6.9:

Dialogue has progressively shifted away from the Suggesting voices towards the Telling voices (but not in their accentuated form as in Phase 1) A conscious effort has been made to do this.

This is a challenging phase requiring directness of interaction while avoiding any inclination to punish for perceived failure.

The principle dynamic movement is between the Asking and the Telling voices with the former anchoring the dialogue. Essentially, we're and to help another person to understand why control has slipped and to coach for self-correction.

In this phase, when Leaders coach, they coach, observe, and listen for conformance and for correction as a prelude to further improvement.

6.6 Developing "Phase 3" Capability

The capability required to make Phase 3 work carries risks and challenges. In Table 6.4 we outline a number of these and from our experience, propose ways of addressing them.

Table 6.4 Risks, Challenges, and Counter Measures (Phase 3)

Risks and Challenges	Countermeasures
Difficulty in combining the Telling and Asking voices	Coaching, skills development from real-time feedback
Avoiding the more 'disciplined' emphasis in Phase 3	Skill-building especially around the 'telling' cluster of verbal strategies, teasing out the differences between them
Not understanding the purpose of Phase 3	Build an understanding of this development theory. Learning to read the situation and apply dialogue in new and appropriate ways. Learning to recognize the impact that 'talk' has on the thinking, feelings, and actions of others
Experiencing the enhanced need for 'control' as a signal to return to pre-CI behavior	As above
Slipping too easily into the accentuated verbal strategies within the Telling cluster of SoundWave	Skill-building especially around the 'controlling' cluster of voices, teasing out the differences between them

6.7 The Specifics of Talk

The voice "to correct" plays a particular role in this endeavor. Good "correction" serves to reinforce the need to adhere to the process providing the foundation for improvement. It is clear and affirmative and works well in conjunction with efforts at coaching for conformance. Our research, however, indicates that it can often be executed poorly and, according to data from the SoundWave self-perception and SoundWave 360 assessments, is manager's least preferred voice.

Developmental work with managers and leaders often has to focus on reducing the use of the "challenging voice" (and accentuated versions of these) and increasing the use of the voice "to correct." This is easily done once people can clearly distinguish one voice from another.

6.7.1 Continuous or "Generative" Improvement: Phase 4—Incremental Step Change

Thousands of improvements every month; hundreds every week; an incremental focus on change; cycles of experimentation structured through PDCA. This is not only the world of sustainable continuous improvement but also the world of generative improvement, a world in which "managers as coaches" leverage the ingenuity of their people.

Every build on the standard, every insight gained from being close to the work that drives a further improvement, "goes beyond the given," generating the possibility for the next step. This is visualized in Figure 6.10. Stephen Spear outlines a number of lessons learned:[3]

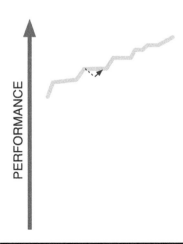

Figure 6.10 Continuous or Generative Improvement.

Lesson 1—There's no substitute for direct observation.
Lesson 2—Proposed changes should always be structured as experiments.
Lesson 3—Workers and Managers should experiment as frequently as possible.
Lesson 4—Managers should coach, not fix.

> The accumulation of many small steps brings benefit in its own right and occasionally leads the way to bigger, 'step changes.'
>
> This is what eons of research have taught us about the nature of learning, about the human mind's capacity to create and to hold gains, namely, that small steps work, (Deming; Pink) and that moreover, small steps harnessed, is a source of competitive advantage.

Characteristically, it is like this ...
We're doing the same thing we always did ... trying every day to improve every little bit and piece. But when 70 years of very small improvements accumulate, they become a revolution.[4]

Typically, the "sound of the site" might be captured in a SoundWave profile shown in Figure 6.11.

The ambition set out on Figure 6.5 of this chapter in the "continuum of conversation" has been realized. Leaders and managers can more fluently

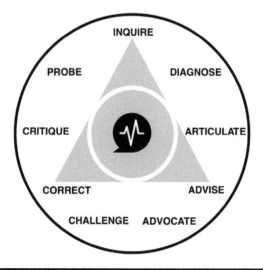

Figure 6.11   SoundWave Profile in Phase 4.

find their way around the SoundWave model as the situation demands it. And yet it is the depth of skill in their coaching role that most allows them to drive the culture of ownership and improvement.

The quality of the conversation is given almost as much importance as the quality of the product, service, or process and in this way, social norms (such as "coaching the coach") embed the required workplace inter-actions institutionalizing them.

When leaders coach, they coach for generative Continuous Improvement, looking and listening for excellence in the quality of the dialogue.

6.8 Developing "Phase 4" Capability

The capability required to make Phase 4 work carries risks and challenges. In Table 6.5 we outline a number of these and from our experience, propose ways of addressing them.

It is from this vantage point that academic and other observers can notice a different quality in the way that leaders interact with their people and people interact with one another and it is from this vantage point that conclusions are made about the "behaviors" necessary for those that follow to target and to emulate.

Table 6.5 Risks, Challenges, and Counter Measures (Phase 4)

Risks and Challenges	Countermeasures
Seeing Phase 3 "stability" as an end in itself	Coach to adopt more coaching skills—ownership more than compliance an outcome.
Holding on to Phase 3 for too long	Looking at the impact on the quality and quantity of improvement actions actually achieved and coaching as above
Not having the breadth of interactional skill developed well enough	Intense 1-2-1 coaching of 'coaching skills'
Viewing the more "coaching" oriented style as discretionary	Institutionalize the required behavior through boss-subordinate feedback on the quality of the dialogue and through the provision of structures that guide the conversation, such as P-D-C-A

6.9 Conclusion—Looking or Listening?

This chapter has argued that progressively shifting the dialogue in line with phases of continuous improvement will contribute to the construction of sustainable, or indeed, "generative" improvement. But can this be true? Can such an achievement be a simple matter of changing the way we talk, of shifting the dialogue?

Well, it's all about the conversation.

Phase 1—When I'm firefighting, I will be (just about) in conversation.
Phase 2—When I'm seeking to persuade others that it's time for a change, I will be in conversation.
Phase 3—When I'm seeking to hold the gains through standardization, I will be in conversation.
Phase 4—When I'm coaching for improvement, I will be in conversation.

It's time we focused more on the *content* of the behavior, on the "how" we deliver the complex social interactions required of things like respect, humility, collaboration, trust, and ownership. None of this is possible unless our face-to-face interactions understand that these are also "mouth to ear" interactions. What I say and how I say it matters. How well I listen to and understand the perspectives of others matters. Combine this with the appropriate use of gesture, with clear alignment between what I say and what I do, and with deep self-awareness and the chances of sustainable operational excellence are high.[5]

Notes

1 By comparison to "body language," our talk is a precision instrument able to specify current reality, conceptualize, and imagine the future and analyze the past. At its best, our huge repertoire of "gesture" can tell us something of the thoughts and feelings of others, but it still requires a considerable amount of interpretation on the part of the receiver whereas "talk" simply spells it out!
2 A comprehensive definition of factors for this level is also set out in the Shingo Prize award level.
3 Stephen J. Spear, "Learning to Lead at Toyota," *Harvard Business Review* (May, 2004). https://hbr.org/2004/05/learning-to-lead-at-toyota (last accessed February 2021).

4 Thomas A. Stewart and Anand P. Raman, "Lesson's from Toyota's Long Drive," *Harvard Business Review* (July–August, 2007). https://hbr.org/2007/07/lessons-from-toyotas-long-drive (last accessed February 2021).
5 Thank you for reading this chapter. Kevin Eyre, Founder and Managing Director, SoundWave Global Ltd, kevineyre@soundwave.global.

Bibliography

Berger, P.L. and Luckmann, T. *The Social Construction of Reality*. London: Penguin, 1966.
Flood, R. *Rethinking the Fifth Discipline*. Abingdon: Routledge, 1999.
Issacs, W. *Dialogue: The Art of Thinking Together*. NY: Random House, 2008.
Rother, M. *Toyota Kata*. NY: McGraw-Hill, 2010.
Senge, P. *The Fifth Discipline: The Art and Practice of the Learning Organization*. London: Century, 1990.
Spear, S. *Learning to Lead at Toyota*. Harvard Business Review, May, 2004.
Watanabe, K. *Lessons from Toyota's Long Drive*. Harvard Business Review, 2007.

Soundwave—Appendix 6.1

The SoundWave model shown in Figure 6.12 and concept describes nine verbal strategies that people use in constructing dialogue. These strategies are valid across cultures and individuals report on their particular and often unique preferences through the use of the SoundWave method which includes assessments such as Brilliance 3, Self-perception, SoundWave 360, and Cultural Assessment.

SoundWave invites its users to reflect consciously on the impact and effect of their dialogue with others helping them to achieve improved relationships and results.

The SoundWave model and concept can be simply categorized into three clusters of verbal strategy. The Asking voices at the apex of the triangle (inquire, probe, diagnose); the Telling voices at the bottom left (correct, critique, challenge); and the Suggesting voices at the bottom right (advise, advocate, articulate). Individuals can measure their "tendency" in the use of the model through the SoundWave Self-perception assessment and the

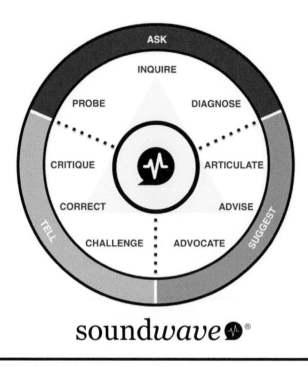

Figure 6.12 The SoundWave Model.

SoundWave 360. Groups can assess their impact and inner dynamic through the use of SoundWave Cultural Assessment. These assessment tools foreground an attempt to develop situational capability in the use of all of the nine verbal strategies.

Chapter 7

Why Bother Seeing Where It Has Worked and the Lesson Learned?

Chapter Summary

While every organization has unique context and requirements there was nevertheless a lot of value to be had from understanding the approaches that other organizations have taken. In this wide range of cases, the authors share in their own words what they did. How it was applied, and the lessons learned. The case studies are from several different sectors and include telecommunications, logistics and distribution, aircraft maintenance, car manufacture, financial services, mining, and food production. The final case explores how maturity assessments have been adapted and undertaken using a virtual approach necessitated by COVID-19 restrictions.

In this chapter, we have asked a range of friends and associates to share their experiences, and we are incredibly grateful for the time and insights they have given us. Each case has been written with a view to sharing lessons learned and has been written by people actively involved in the work described.

7.1 NBN Co. Case Study

Indrajit Ray, Clyde Livingston, and Richard Perry

Introduction to the Organization and the Type of Assessment

NBN Co. was established in 2009 to design, build, and operate Australia's wholesale broadband access network. Underpinned by a purpose to connect and lift the digital capability of Australia, NBN Co.'s key objective is to ensure all Australians have access to fast broadband at affordable prices and least cost. As an Australian government-owned enterprise, NBN Co. set about planning and delivering this major infrastructure project and mobilized to operate the resulting national asset.

Why Assessments Were Important and How They Were to be Used

In 2014, newly appointed NBN Co. CEO, Bill Morrow, introduced a new focus for the company with effective process management at its core. Using process management, key gaps were identified and addressed. Roles and accountabilities became clearer, and leaders were now held much more accountable to deliver Key Performance Indicators (KPIs) and the means to improve them.

To understand how well this was being lived day to day, process maturity assessments were introduced to effectively assess teams and their leaders on how well they understood their processes, and how well they were working both upstream and downstream to improve them for the benefit of customers. An excerpt from the process maturity playbook is shown in Figure 7.1.1.

Core, value-chain processes (delivery of value to customers) were set as a target of maturity at Level 4, and all other supporting processes set a

> For most organizations there is a significant gap between their aspirations for higher levels of business performance and the lived reality.
>
> At NBN Co, we regularly assess the maturity of all Level 2 processes to understand the gap between where we are now and where we need to be and action the best possible solution to close the gap.
>
> Our process maturity assessment model guides process owners and teams through an evolutionary roadmap and helps them turn the BPE (Business Process Excellence) principles into plans and actions that will drive the required performance uplift.
>
> In FY20 and beyond, we will execute our company strategy through continually optimized, and actively managed, end to end business processes.
>
> (Excerpt from the NBN Co Process Maturity Playbook)

Figure 7.1.1 Excerpt from NBN Co. Process Maturity Playbook.

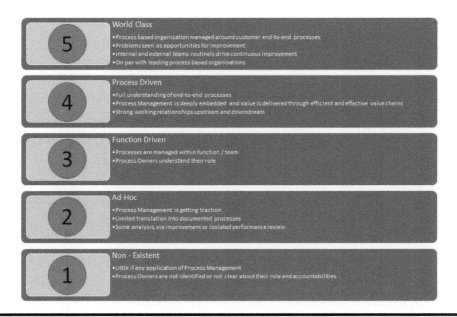

Figure 7.1.2 Summary Descriptors for Process Maturity Levels.

target of maturity at Level 3. The summary descriptors for each level are shown in Figure 7.1.2.

The Sponsor of Assessments and How They Sold the Why

The process maturity assessments were sponsored by the CEO and strongly endorsed across the entire C-suite management levels. This needed to be the case, as a close collaboration was required across HR, Strategy, Delivery, and Operational teams.

The organization needed to double its rate of connections (connecting homes and businesses to the new network) for four consecutive years.

The "sell" of process maturity assessments from the C-suite to their teams was based on "if we don't embrace this methodology and work as a single team to quickly identify and resolve issues—we will never hit our targets."

The Desired Behaviors Being Measured and Assessed

There were six sections to the assessment, with around four specific behaviors being evaluated in each. The process owner would typically

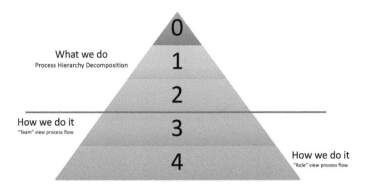

Figure 7.1.3 Process Hierarchy.

provide the majority of the responses. It is important to look for the three elements of toolset use (i.e., the process modelling tool); skillset (i.e., the capability to conduct process improvement); and, most critically, mindset (i.e., the belief that value will come from this work). The six sections were:

1. Prove that your processes are documented, used, and maintained by your team, and linked upstream and downstream.
2. Prove that your processes have clear accountabilities and improvement metrics and that these accountabilities are allocated to your team.
3. Demonstrate that upstream/downstream metrics (at hierarchy Level 3—see Figure 7.1.3) are linked to your processes and that you meet with "adjacent" process owners to improve the outcome of the overall flow.
4. Show how your team reviews these metrics and takes ownership of improving them.
5. Showcase recent improvement activities and outcomes—best done by the team involved rather than the process owner—to test how deep the mindset is embedded.
6. Prove that risks are being managed in your process; if there is a separate enterprise risk-management program in place, this may not be required.

The Design of the Assessment System

The design of the assessment system varied between critical (core value stream) and non-critical processes (everything else). For critical processes, the central team conducts one self-assessment and one formal assessment per annum, roughly six months apart. Non-critical processes are self-assessed once a year but a formal assessment can be requested at any time.

To maintain a level of currency, the following tactics were used:

An annual assessment calendar was published and circulated on internal systems.
A C-suite member would send the communications to their process owners asking for their participation and support.
The criteria of expectations, checklists, templates, maturity charts, examples, etc. all published on the company intranet to help process owners as much as possible.
Assessment results were reported up to C-suite level.
A link was established between the process maturity score and the annual bonus remuneration for process owners.

The assessment usually consisted of meeting with the process owner to get the first view of all the elements being tested, then subsequently (and most importantly) with the team/s for them to demonstrate the same. Any variations between the two sessions were usually worth exploring to understand where the misalignment came from. This was often a surprise for both process owners and their teams and formed part of the action plan that followed the assessment.

Alignment of Assessments, Assessors, and Calibration, Including Development of Internal Assessors

From 2015 to 2018, a single, central assessor led formal assessments for all critical processes. This role ran an induction program for secondary assessors to ensure a common understanding and interpretation of the maturity criteria. This ensured each assessment involved a supporting independent assessor from another business unit and subsequently facilitated calibration sessions to ensure consistency and integrity of assessment results.

In 2019, multiple lead assessors led the assessments. Efforts were made to allocate a single lead assessor for each business unit, so a consistent approach could be taken for core parts of the value chain. Detailed guidelines and checklists (see Figure 7.1.4) were published when a larger cohort of secondary assessors was inducted. These guidelines and checklists were leveraged successfully as business stakeholders become more involved in assessments in 2020.

164 ■ *Why Bother?*

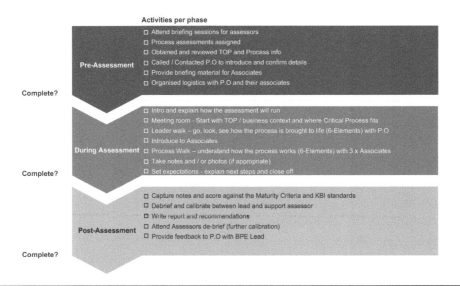

Figure 7.1.4 Assessor Checklist.

Selling the Results—Expectations and Use of the Assessment Results

Assessors presented a detailed report—with observations, gaps, and what was working well—back to process owners and supporting Business Process Excellence specialists. Process owners then formulated a BPE plan to uplift process maturity and to address the gaps found in the assessment. Typically, the supporting BPE specialists worked with the business teams to implement this plan and to prepare the process area for assessment in the next cycle.

The central BPE team then produced an executive summary that highlighted trends from the process maturity scores, broken down by business area and process area. A comparison was also made on the previous year's process maturity. This was accompanied by commentary, often pointing out key business changes or evolving assessment criteria.

The executive summary was presented to NBN Co.'s Chief Strategy Officer and shared with the C-suite. This enterprise view on process maturity provided executives and leaders within NBN Co. with the depth of control, collaboration, and alignment required to facilitate and drive target business performance. An example extract is shown in Figure 7.1.5.

To ensure **nbn** is able to execute the strategy through **continually optimized, and actively managed business processes**

L2 Critical process targets
- **Critical process to M4**
- **Non-critical Processes to M3**

Twice yearly assessment of all L2 processes to **understand the gap** and develop a plans to **close the gap**

Key Artefact:
1. Process maturity Criteria
2. Process Maturity Playbook
3. Assessment process
4. Bringing nbn values to life

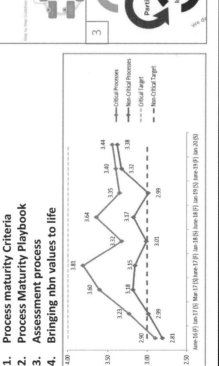

Figure 7.1.5   Sample Report Extract.

Focus on Role Modelling and Best Practice Sharing

Process owners and business teams that had successfully matured their processes were invited to showcase their success in intranet articles and "bright spot" videos. They were shown at team meetings and departmental town halls, highlighting team behaviors that uplifted process maturity and drove business performance.

The Deployment Plan for the Assessment Program and for Developing Process Maturity

Process owners in each business unit were supported by BPE specialists who helped them prepare for process maturity by facilitating artifact creation, delivering business improvement, and leading transformation. The same BPE specialists formed a BPE deployment plan based on recommendations contained within the assessment report.

Typical BPE deployments, post-assessment, involved:

- Supporting process owners to walk through their team's processes to uncover where issues were arising.
- Reinforcing team operational habits to maintain business performance (e.g., visual management boards and huddles).
- Bolstering the collaboration and alignment of key metrics with upstream/downstream business processes.
- Selecting the appropriate problem-solving approaches that would drive the team's continuous improvement and innovation program (e.g., Just Do It vs PDCA vs DMAIC vs Design Thinking, etc.).
- Identifying risk at the process level and implementing process controls as continuous improvement opportunities.

The Next Steps in Maturing the Program

The ambition for the program is to become self-sufficient in the organization and to become part of the culture—"the way it works around here."

Process Owners will not only assess their own processes, but also ultimately replace central assessors to ensure optimal control, collaboration, and alignment across core value chains—to support the transition to a customer-led service organization.

Lessons Learned

Several factors proved to be influential in making this a success for NBN Co.:

C-suite level sponsorship
a massive scale challenge to achieve
belief that mindset is more important than skillset and toolset (but you need all three to nail it)
resources made available to run and sustain the program over a number of years

Going forward, some of these factors will undoubtedly change. The focus for the executive team has already changed and will continue to evolve, so, "How do we keep high-level sponsorship?" The massive scale challenge has recently been achieved, so, "Do we need this capability anymore?", and so on.

Regardless of what happens, NBN Co. has achieved an incredible goal which has been publicly credited to the capability of the teams and the way they solved the problems and delivered the vision. Process maturity played its part in this and is embedded in the culture and language of the business.

Indrajit Ray, Senior Manager Process Transformation, nbn co. ltd (nbn), has contributed to the text in this chapter with nbn's permission, however nbn makes no warranties as to the accuracy of this text and no reliance should be placed on it.

7.2 Panalpina Logistics Case Study

Andrew Lahy, Maria Pia Caraccia, and Mike Wilson

With thanks to the Panalpina logistics team who developed, improved, and implemented LogEx across the world.

Looking at rolling out an assessment system across the whole enterprise has both great strengths in sharing best practices and business understanding and great challenges in having only one system that works for technical operations and support functions alike.

This case study explains how the Panalpina logistics division turned around annual losses of $40 million into an annual profit of $10 million per year. The case study explains the journey, good and bad, as their logistics division turned from being a loss-making, ugly-duck division in 2010 to a Shingo medallion winning team in 2019. The case study explains how and why the company started to use assessment tools and how the tools evolved over the nine years. The team provides their top tips on what worked for them and the lessons learned from their journey.

A Turning Point for the Company and the Start of the Operational Excellence Program in Panalpina Logistics

The year 2010 was a turning point for Panalpina. It was the year the company finalized the payment of over US $80 million to conclude an investigation by the United States Department of Justice into bribery claims. With that chapter closed, Panalpina launched a new strategy to return the one-hundred-year-old company back to profitability and growth.

At the heart of Panalpina's new strategy was the idea to transform the company's logistics division. At the time, Panalpina was well known for its international air and ocean freight services and was one of the market leaders in this area. But, unlike many of its competitors, Panalpina had resisted providing warehousing and logistics services and did not actively sell these services to its customers. The new strategy aimed to change this and transform Panalpina from a specialist international freight business into a wider supply chain company.

In 2011 Panalpina hired Mike Wilson, a seasoned executive who had held senior positions in both large manufacturing and large logistics companies, and asked him to lead the transformation of Panalpina's logistics division. "When I arrived in 2011," Mike explains,

> Panalpina was already providing some logistics services but logistics was seen as the ugly duckling of the company. Most people in the company saw it as a necessary evil, something the company had to provide to keep its international freight customers happy, but not something that was a core offering. Logistics was often treated as a cost to be minimized rather than a value-adding service that could be provided to customers to generate profits.

Mike knew he had to change that thinking.

Mike set about building a team to transform the Panalpina logistics business and as a former student at Cardiff Business School, well known for educating many senior Lean leaders, he knew the power of Lean and Operational Excellence and how these could be used to change the culture of a team and a company. Mike further says—

> I've always believed that Lean, Operational Excellence, Total Quality Management, or whatever you want to call it—the name is irrelevant to me—is quite simply about engaging and involving people in the business and allowing everyone to play their part in making regular incremental improvements. It really is as simple as that. It is certainly not about complex tools, certificates, or assessments. In fact, complexity is often the enemy. The key is creating a program that is open to everyone in the business, not just to a small group of experts. I have always believed that the key to creating a sustainable operational excellence program is to allow the operations teams themselves to lead and develop the program, not force anything on people from the top down.

With this in mind, Mike asked a small team to see how a new program could be created to involve and engage everyone in the logistics business in continuous improvement and operational excellence. The team decided that if they wanted to design a bottom-up operational excellence program, the program had to start with the involvement and input from front-line operations teams. So that is exactly where the Panalpina logistics operational excellence journey began.

Creating the First Operational Excellence Assessment Tool

The transformation started by working with the operations team to put on paper what made a good warehouse operation successful. What were the key components that enabled a warehouse operation to run well? Through a series of workshops and iterations with warehouse operation teams in the Netherlands, the United States of America, Singapore, and Dubai, the team developed a pyramid of excellence, shown in Figure 7.2.1.[1]

The team defined the bottom two layers of the pyramid as the basic components that all good warehouses should have in place. These bottom two layers were defined as the Bronze level in the pyramid and the pyramid continued upward with Silver and Gold levels. Behind each block in the pyramid were four questions that a site could answer with a simple yes or no response to check that the basics were in place at their site.

170 ■ *Why Bother?*

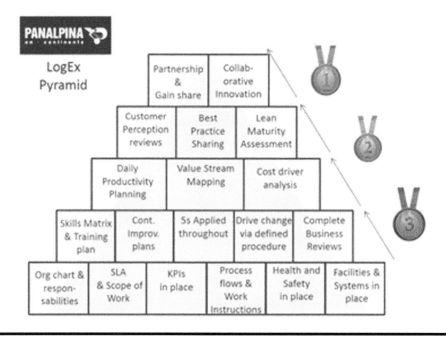

Figure 7.2.1   LogEx Pyramid.

Although it was not really the intention at the start, the team had developed a first, simple assessment tool that sites could use to check that they had the basics in place. The assessment tool was created as a simple Excel sheet and, rather than being enforced from the top or setting up central assessment teams, sites were encouraged to carry out the assessment themselves and just communicate when they had all the basics in place. Although not mandatory, many sites took the opportunity to use the assessment tool to demonstrate that they had the basics in place. Then, rather than criticizing those sites that had not used the tool, the team took the decision to congratulate those sites that did. In this way, the original pyramid was not created as a head-office assessment tool, but rather a self-assessment tool that sites could optionally use to demonstrate where they were on the pyramid.

This first pyramid tool became known as the LogEx (Logistics Excellence) Pyramid, and the name LogEx eventually became the name of the wider operational excellence program at Panalpina. Although not the intention at the start, the pyramid became a strong visual symbol of Panalpina's operational excellence program.

But, despite some good early momentum and a simple tool to start the program, the limitations of the LogEx pyramid soon became apparent. On one side, those sites that had used the pyramid to get the Bronze basics in place then wanted to push to the next level of Silver in the pyramid. These higher levels of Silver and Gold had not been fully defined nor fully thought through. On the other side, as the assessment was optional, there were still several sites that showed no interest in using the pyramid and getting involved in the new LogEx program.

It was at this time that the team sought some external help from Professor Peter Hines, a specialist in creating long-term sustainable operational excellence programs.

Problems with the LogEx Pyramid: From Tools to Strategy

"We were quite proud of the pyramid. We had been using it for about twelve months and it had proven to be quite popular in the company and we thought we were on the right track," explains Andrew Lahy, who was part of the team that helped to set up the pyramid. "When we proudly showed the pyramid to Peter, he was quite dismissive and said we had done what many others had done before—created a tool, but not created a strategy behind it to make it sustainable." The Panalpina team decided to take Peter's advice and began to introduce the Lean Business Model (see Figure 7.2.2) to the sites that were keen to move beyond the Bronze basics of the LogEx Pyramid. With this, the LogEx Pyramid, rather than being at the heart of Panalpina's operational excellence program, became just one of the tools and techniques within the wider Lean Business Model shown in Figure 7.2.2.[2]

The introduction of the Lean Business Model had a mixed response from the wider Panalpina team. For some sites, particularly those who had been working hard on the next level in the pyramid, a move away from the pyramid towards the new Lean Business Model was seen as "moving the goalposts." Andrew Lahy explains,

> On one side, we felt bad, as the sites were right, we had moved the goalposts. When we created the first pyramid, our focus was on the Bronze elements of the pyramid and we had not really put much thought into the Silver and Gold elements. Once we had listened to Peter, though, and understood the importance of linking the tool to a wider operational excellence strategy, we realized that we had to move beyond the pyramid to keep the program going and push ourselves to the next level.

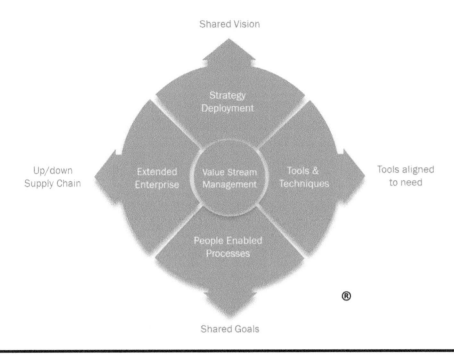

Figure 7.2.2   The Lean Business Model© S A Partners.

To compromise, the team kept the original pyramid as the basis of the assessment system but incorporated components of the Lean Business Model into the pyramid so that those sites who wanted to get to the Silver level also had to demonstrate the five key components of the Lean Business Model: a strategy, value streams maps, people engagement, tools linked to the strategy, and a partnership approach with customers and suppliers.

This way, the pyramid, which was slowly becoming a recognized icon in Panalpina, could stay, but it now incorporated elements of the Lean Business Model and provided a way for those sites that had moved beyond the Bronze level to start assessing themselves against the wider Lean Business Model.

Problems with the Pyramid, Again: From Strategy to People and Behaviors

In 2016, Maria Pia Caraccia took over the running of the LogEx program at Panalpina, and although she had used the Lean Business Model and liked it, she also saw its limitations. As Maria Pia explains,

As soon as we started talking about strategy, this was perceived as something that management does and is responsible for. The risk for us was if we started putting too much focus on strategy, we would lose the interest of front-line operations teams, and it was these teams that had always been the drivers and key to the success of our operational excellence program thus far.

Maria Pia decided to explore different approaches which focused more on the people side of things and made the decision to include elements of Shingo assessment into the LogEx Pyramid. As Pia explains, "When we started to introduce some ideas from the Shingo model into our assessment, we had the same issue again in that some sites complained we were moving the goalposts." This time though, rather than feel bad about introducing some new elements, Pia explained to the teams that to continue to push the bar higher, it was also necessary to constantly improve the continuous improvement approach, and this included improving the assessment tools being used in Panalpina.

Pia now added components of the Shingo assessment into the Gold stages of the original LogEx Pyramid. As Pia explains, "We decided to go for the Shingo award, not because getting the award in itself was important, but it allowed us to benchmark where we were compared to the very best companies in the world." Although the team had not planned it at the start, they had moved from creating an operational assessment tool in the first iteration of the pyramid to an assessment incorporating strategy and direction in the second iteration, and finally to one focused on behaviors and values in the third iteration. The transition through the three types of assessments is summarized in Figure 7.2.3.

As Pia explains,

> The Shingo assessment really did challenge our teams, as it required that everyone in the organization could not only articulate the right behaviors needed to drive sustainable continuous improvement, but also could demonstrate those behaviors in practice over a long time frame, and by long, the Shingo assessment looks for sustained continuous improvement over at least a five-year period, ideally more.
>
> What we learned from assessing ourselves against the Shingo criteria is that, although we were on the right track and now could proudly compare ourselves to some of the best companies in the world in

174 ■ *Why Bother?*

Phase 1: Development of the LogEx Operational assessment tool

Phase 2: Using the Lean Business Model to link to strategy and direction

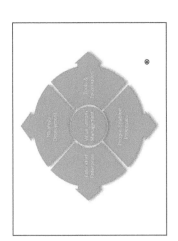

Phase 3: Using the Shingo model to link individual behaviours into the approach

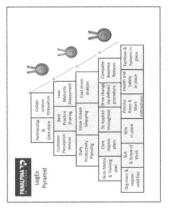

The LogEx pyramid assessment, developed internally by the operations team, allowed us to get started and ensured the first assessment were simple and understood by everyone.

The downside was that the pyramid had no direct link to the wider company strategy.

The Lean Business Model assessment allowed us to link the LogEx pyramid assessment to the wider company strategy.

The downside was that once we started talking about strategy, this was perceived as "management" talk and something decided in the office. We needed a way to create a bridge between the Lean Business Model and the original, operationally developed LogEx pyramid.

The Shingo model assessment challenged everyone in the business to think beyond strategy and tools, and think more broadly about how individual behaviours, regardless of your role in the company, influence the direction of the business.

The downside is that working towards a Shingo Prize cannot be done by just a few people in the business over a short period of time. It really is a long-term investment and requires a full commitment from everyone to make it work.

Figure 7.2.3 The Transition from the LogEx Pyramid, to the Lean Business Model, to the Shingo Enterprise Excellence Model.

terms of operational excellence, our journey, even nine years in, was only just beginning.

One thing was for sure, though—by 2019, when Panalpina was awarded a Shingo medallion for operational excellence, logistics was no longer the ugly duckling of the group. It was generating a profit and setting the standard in the company and the industry for sustainable continuous improvement.

Our Top Tips and Lessons Learned

1. **To get started with assessment tools, use something that is created by people in your company**. You and your team know better than anyone what your business needs, so copying an assessment from an outside company is unlikely to work. What is more, by involving different people in the business in designing the assessment you can make sure the language and terms used are relevant and understood by everyone.
2. **Do not let the assessment tool become the objective for improvement.** As Maria Pia explains, "Getting Bronze, Silver, or Gold level in the LogEx Pyramid had no direct benefits for our customers, our employees, or shareholders, so if teams were just focused on getting to these levels in the pyramid they were invariably focused on the wrong things. The sites that really excelled were those that set their own strategies and plans and used the LogEx Pyramid as a framework to support their plans. The assessment was just a means, not an end."
3. **Do not over plan your operational excellence journey**. As Andrew Lahy explains, "One thing I think we did get right from the start is that we just got started. I think if we had presented a big, long-term road map and business case, we would have spent months discussing the long-term plans at head office. Instead, we kept it small and simple, and just got started doing the assessments at our sites. Once we had some momentum and success, then we were able to start thinking about strategy and long-term plans. I would love to say this was due to some deep insight and knowledge on our part, but it was not. We just got lucky that it worked."
4. **Do not be afraid to change your assessment criteria.** Almost within twelve months of rolling out the first LogEx Pyramid, we could see there were some issues, particularly with the blocks we had

defined in the Silver and Gold categories. One of the most fundamental problems was using a pyramid shape—we wanted the elements for Silver and Gold to be harder and broader than the elements in Bronze, but our pyramid was narrower at the top than the bottom! We thought about changing the shape or even making it into a 3D pyramid, but in the end, we just worked around this limitation. The key lesson for us was that you cannot keep changing the assessment criteria too often, but you cannot be afraid to change it eventually, as what the business requires will change over time and your assessment needs to change to reflect this.

5. **Have some fun with the assessment.** Often managers and teams already have enough audits, KPIs, and objectives to work on. If the assessment is seen as just another piece of bureaucracy or a means for head-office teams to come in and audit operations and find faults, then it is unlikely to get traction or the buy-in from site teams. We learned that sites needed to see the value and benefit of using the assessment themselves and find creative ways to engage their teams and involve them in continuous improvement. Our personal favorite was in the Netherlands, where the management team had a party and brought a cake to celebrate being the first site in the world to achieve Bronze.

The Business Benefits

One of the most common questions we were asked when we introduced the LogEx approach was, "But what profit will it bring?" Or, the alternative question, "But what is the expected ROI on this project?"

Our simple answer was and is—none. There is no clear ROI calculation for an operational excellence improvement program. Introducing an operational excellence approach or assessment tool will not magically improve profit. In fact, at worst, introducing new tools and assessments can be a distraction from the core purpose of the business, which is for a team of people to provide a service or product to customers. So, for us, we did not introduce LogEx based on an ROI calculation or profit expectation. We did it because we wanted to make Panalpina a great place to work and create a framework to allow everyone in the company to be involved in making improvements every day. We knew that if we could get all our teams working together to improve the business, in the end, this would translate into a better and higher quality service to our customers. Doing this does not guarantee you will make a profit, but not doing it will guarantee you will go out of business! So, for us, the introduction of LogEx and the assessments was just one of the ways to help us harness the collective brains and skills of all the team.

We found that when you have a group of employees who actively want to make improvements and they have the framework, support, and tools to make them, then the profit will come as a result. It is a natural by-product of a successful operational program.

Although we did not set up the LogEx program based on a profit target, looking back, we achieved one. When we started the program, the logistics division was losing US$40 million. After nine years, the division was making a US$10 million profit. Was this turnaround just down to the LogEx approach and assessment tools? Absolutely not. LogEx was just one of the components. The real improvement came from the Panalpina logistics team, who made the improvements every day over the nine-year period.

7.3 Airbus Australia Pacific Transformation Program—Behavior Assessments Case Study

Kim Gallant

Airbus Australia Pacific (Airbus (AP)) is a leading aerospace company that designs, manufactures, and delivers industry-leading commercial and military aircraft and helicopters, satellites, and launch vehicles, as well as data services, navigation, secure communication, urban mobility, and other solutions for their customers on a global scale.

This case study covers a period when Airbus (AP) faced declining competitiveness and market share. The case study describes the approach taken by Airbus (AP) to improve customer satisfaction, product quality, safety, employee engagement, and, ultimately, competitiveness. While Airbus (AP) had a long history of applying Lean principles, these were traditionally confined to its manufacturing and supply divisions.

The Airbus (AP) transformation commenced in 2016 with the introduction of the Airbus (AP) Operating System (AOS) and New Ways of Working. The AOS and New Ways applied Lean principles that transformed the system of improvement to the entire operations and moved Airbus (AP) from initiative-based, individual projects towards a culture of continuous improvement, where every employee engaged in the daily improvement of the Airbus (AP) business. The case study explains why and how Airbus

(AP) used a system of assessment to ensure management teams could track progress against the Airbus (AP) strategic objectives and their annual plans.

Beating the Transformation Odds

To drive fundamental culture change, the Airbus (AP) executive committee released Strategy 2021, which incorporated three key cultural change initiatives:

SI 1. Focus everyone on customer value to deliver product and service excellence.
SI 2. Secure and develop people with the right skills knowledge and attitude to realize our vision.
SI 3. Foster a sustainable Lean culture (with a Lean maturity target of Level 2 by 2015 and Level 3 by 2016).

But it was not as simple as launching the transformation and then hoping for the best. There were two main problems to overcome.

1. Even in Australia and New Zealand, Airbus (AP) was a diverse business with multiple programs involved in the transformation, in various stages of their life cycle. How could they set meaningful and consistent targets across these programs and sites?
2. Additionally, experience within Airbus (AP) and recent industry surveys showed that only two percent of companies that introduced a transformation program based on Lean principles achieved their anticipated results. Other studies found that many transformation programs had not sustained their progress after initial, short-term success.

Further reading by the Airbus (AP) transformation team showed that transformation programs that incorporated some form of maturity assessment to monitor results, assess gaps, and then provide insights on how to close them proved the most effective when it came to delivering on intended transformation outcomes.

The diagram in Figure 7.3.1 shows that the transformational results were built on a focus on its people and the creation of a Lean culture and capability across the company. Airbus (AP) also made both contractual and public commitments to our customers to deliver improved benefits

Figure 7.3.1 The Target of the Airbus Australia Pacific Transformation.

via an embedded culture of Lean thinking and continuous improvement. This culture would ensure the efficient flow of value and delivery of improvements that were important to Airbus (AP) customers. Simply put, this meant making it easier for Airbus (AP) people to do their jobs and for Airbus (AP) customers to do business with us.

Figure 7.3.2 shows the principles behind the New Ways of Working at Airbus (AP). ***My normal work conditions*** provided clear and transparent visual performance management so that everybody understood how their contribution counted. ***My visual alerts*** ensured that performance was reviewed at the shortest possible intervals of control so that smaller problems could be solved at the team level more often. ***My reactions*** engaged and aligned team members to the company goals and vision.

Finding the Right Sponsor and Positioning the Transformation for Success

Applying Lean to the transformation of Airbus (AP) was a strategic decision. Explicitly linking Lean in this way ensured that it was sponsored by the executive committee and flowed down through Management by Objective targets to all Airbus (AP) teams.

The transformation of Airbus (AP) was clearly focused on their customers. The executive committee communicated the results of the

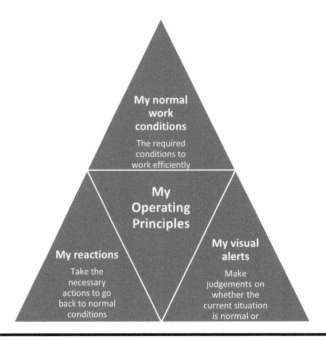

Figure 7.3.2 Continuous Improvement Principles at Airbus Australia Pacific.

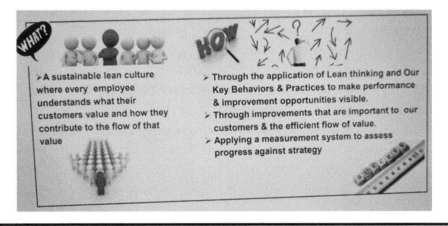

Figure 7.3.3 Applying a Measurement System to Ensure Progress Against Strategy.

Voice of the Customer (VoC) through planned patterns of communication that ensured every employee understood each customer's underlying needs. They also made certain that every employee understood the intent of creating a Lean culture, and how the system of assessment would be used to ensure progress against the strategy and could be measured (Figure 7.3.3).

A New Way of Measuring Success

The leadership team understood that a key element of this transformation's success was going to be a change in employee behavior and this needed to be measured in some way to gauge progress. Airbus (AP) defined five key management behaviors and five key practices.

Management Behaviors:
service mindedness
welcome problems
accountability
rigor and follow standards
continuous improvement

Management Practices:
PDCA
visual management
practical problem solving
Gemba walks
customer feedback loops

In support of these management behaviors and practices, a behavioral framework was developed and embedded into human resource documentation, such as position and role descriptions. It was against this framework that annual targets were set and the sustained maturity assessment program used to measure the organization's maturity.

Designing the Assessment Program

Airbus (AP) maturity assessments were initiated shortly after the transformation started and an off-the-shelf approach was selected to ensure assessments could be started immediately. These assessments were conducted against the five elements of the SA Partners Lean Business Model (Figure 7.3.4) and the Airbus (AP) behavioral framework. The inclusion of the behavioral framework in the assessment approach made sure Airbus (AP) teams understood that how results were delivered was just as important as the results themselves.

The assessment system comprised a high-level policy document that established why assessments were important to the business, when they

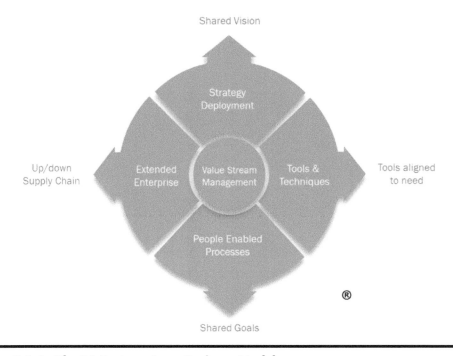

Figure 7.3.4 The SA Partners Lean Business Model.

would be completed, and who would complete the assessments. Originally, the policy defined an assessment period of two years, with this timing selected as it fell at the midpoint of the five-year strategic plan cycle and would present an opportunity to assess how the strategy deployment was progressing. However, the reality was quite different, as the operations management teams could see the value these assessments were providing to their planning and consequently began asking for annual assessments as inputs into the next year's integrated operations plan.

The policy also included an entry requirement that each leadership team had to satisfy before undergoing their first baseline maturity assessment. This ensured all Airbus (AP) teams started their transformation journey in a consistent manner. This policy was supported by a detailed procedural document (Figure 7.3.5) that defined how assessments were completed.

Selecting the Right Assessors

Assessors were selected from the Airbus (AP) management teams and underwent external assessment training and coaching. To build their competencies, they then participated in several baseline assessments

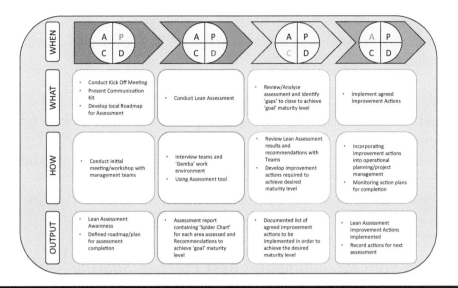

Figure 7.3.5 Extract from the Lean Maturity Procedure.

with an experienced external assessor before participating in the maturity assessment program as an internal assessor.

The lead assessors were selected from this cohort of internal assessors and they completed additional Shingo training to further develop their understanding of the importance of behaviors within a system of working and improving.

Selling the Assessment Results

The results were used as a key input into the operations management team's next integrated operations plan, with the assessment seen as a crucial step in showing progress against both the five-year transformation plan and the annual plan. Teams understood that this progress was not just about organizational maturity, but also an important indicator of readiness for further transformational change, including technological change.

Early adopters of the New Ways of Working who could show a deep understanding of their customer's needs, had adapted their value streams to deliver that value, developed their people, and demonstrated the behaviors that ensured this change was sustainable were given access to the latest technology made available by the digitization of Airbus (AP).

The outcomes of an assessment were communicated directly to the management team, who underwent assessment in the form of a written report that included:

- Strengths and Opportunities identified during the maturity assessment.
- A visual representation of progress against strategic targets for each area assessed.
- A heat map showing where every employee's current behaviors fell within the Airbus (AP) behavioral framework.
- The consolidated answers to a catch-all question aimed at defining what "must" be changed to better engage Airbus (AP) people.

Leader-led Role Modelling and Best Practice Sharing

As part of the New Ways of Working program, every leader within Airbus (AP) went through an eight-month training program to learn how to coach and develop their teams. Annual pavilions (exhibitions) showcasing the improvement work that teams had delivered over the previous twelve months were held across the Airbus (AP). Team members involved in delivering these sustainable changes for the better presented their work to their international colleagues, discussed lessons learned, and simplified paths to replicate their results.

Key Learnings from the Airbus (AP) Assessment Program

Once you have selected a credible approach to assessment, apply it consistently across the business. Airbus (AP) selected the SA Partners Business Model and Airbus (AP) behavioral framework as the basis for their maturity assessments. An experienced SA Partners assessor conducted all the initial maturity assessments to establish a clear baseline from which to measure transformation progress and success. All operations and support teams across Airbus (AP) underwent an initial baseline maturity assessment.

Do not be afraid to adapt your approach to assessment, particularly if the business finds value in the new approach. The original assessment policy in support of the Airbus (AP) transformation defined a biennial assessment period with the customers of the assessments being both the operations management teams and the executive committee. The approach was adapted to facilitate assessments both annually and once every two years after a continued pull from the business to shorten the assessment cycle. Management teams found two key elements of value in annual assessments. The first involved having an external "look-in" to their program that identified issues they were blind to, and the second was that the outputs of the annual maturity assessments improved the quality of the following year's annual integrated operations plan.

Maturity assessments are not audits. It is natural for management teams new to maturity assessments to initially see them as audits, particularly in industries with high levels of regulatory oversight and reliance on an active quality management system. The danger for maturity assessments in these environments is that management teams take the outputs of the maturity assessment and treat them as non-conformances. This drives a negative connotation to the assessment, rather than the desired positive approach where assessment outcomes are seen as opportunities to drive improvement. It will take high levels of communication at the commencement of the assessment program to alert the business to the differences between assessments and audits. Another way to avoid this negative perception of maturity assessments is to carefully word the opportunities for improvement, preferably using the words of the business team being assessed who are suffering the poor process.

7.4 Commonwealth Bank of Australia Productivity Program—Behavior Assessments Case Study

Morgan Jones

Introduction to Organization and the Type of Assessment

Commonwealth Bank of Australia (CBA) is the largest bank in Australia and forty-fifth largest in the world. It engages in the provision of banking and financial services. It operates through the following segments: Retail Banking Services, Business and Private Banking, Institutional Banking and Markets, Wealth Management, New Zealand, Bankwest, and International Financial Services and Other. The Retail Banking Services segment provides home loans, consumer finance, and retail deposit products and services to all retail bank customers and non-relationship managed small business customers.

In a ten-year period from 2003, CBA implemented two formal CI programs. The first, called Commway, focused on the Six Sigma methodology and was used for the first three to four years. CBA then reinvented and relaunched this program, as 'Productivity,' which it used for a further six years. This new program included Process Excellence and Business Process Improvement. Both programs developed short-term hard savings through training a selected few practitioners to execute the improvement but with little sustainability.

In 2013 the new CEO, who was a great supporter of continuous improvement, initiated a new CI program at CBA. The new program developed a 3C model for the productivity program. Figure 7.4.1 shows a schematic of the 3Cs Productivity Program that was implemented.

- Capacity is the savings/efficiency improvement that is being generated and verified.
- Capability is the building of a strong expertise in CI through a structured training and certification program, including yellow, green, black, and master black belts.
- Culture is building the basic CI capability in all employees through specific habits. This is documented very well in the book *4+1*.[3]

This case study will focus on the implementation of the accreditation program implemented to measure the maturity of the culture pillar of the 3Cs model.

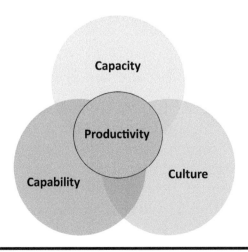

Figure 7.4.1 The 3Cs Productivity Program.

Explain Why Assessments are Important and How They were Used, Including Prerequisites and Maturity Progression in Order to Improve

To drive fundamental culture change, the Productivity team at CBA developed an accreditation program with the goal of assessing how well productivity mindsets and behaviors are embedded across each business unit—comparing it against best practice elements. Accreditation is awarded at three incremental levels: Bronze, Silver, and Gold levels.

Awarding accreditation levels started to drive a desire at a team or department level to continually build a culture of CI. The true value of maturing the CI culture through the accreditation program was reaped well before accreditation was awarded. The accreditation program recognized the commitment each business unit made to:

building productivity capability and culture
understanding the customer and their needs
implementing numerous improvements that brought tangible benefits to both the customers and the team

The accreditation program recognized which teams were more mature by consistently focusing on making tomorrow better than yesterday. Figure 7.4.2 shows the graphic used in communicating the accreditation program.

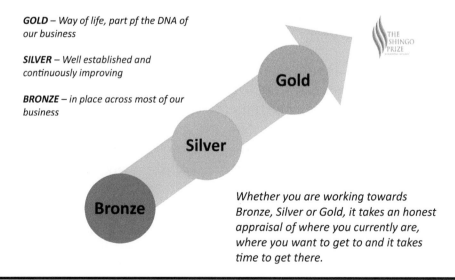

Figure 7.4.2 The CBA Accreditation System.

188 ■ *Why Bother?*

Table 7.4.1 Maturity Levels of the CBA Accreditation Program

Level	Bronze	Silver	Gold
Assessment	Self-Assessment independently reviewed	Center team independent assessment	Expert external plus center team independent assessment
Frequency of Assessment	Minimum annually, most moved to 6 monthly	Annually	2 yearly
Pre-requisite	None	Bronze	Maintained Silver

As teams progressed through the accreditation levels, there was an aspiration that the final goal was that teams could apply and challenge for the Shingo Institute Prize for Excellence. Table 7.4.1 illustrates this journey.

The initial four CBA productivity habits were:

visual management boards (VMBs)
huddles or team stand up meetings
continuous improvement (CI)
standard operating procedures (SOPs)

Figure 7.4.3 is a graphical representation of the core habits working as a system.

Who was the Sponsor of Assessments and Where They Sat in the Organization, Including How They Sold the Why

Applying the accreditation program became a strategic decision, explicitly linking productivity results with individual capability and the culture built into the strategic pillar. The commitment to follow an accreditation program was included in the company's reporting to stock market analysts. This gave the culture pillar of the 3C model high importance, as it informed the market that CBA was building sustainability on savings and improving customer focus.

The CEO was the overall sponsor of the program and he delegated the implementation of the program to the Executive General Manager of the Productivity program. However, to drive accountability and ownership at the executive level there were compulsory targets set for the culture and capacity pillars. There was a hard dollar saving set with a stretch target

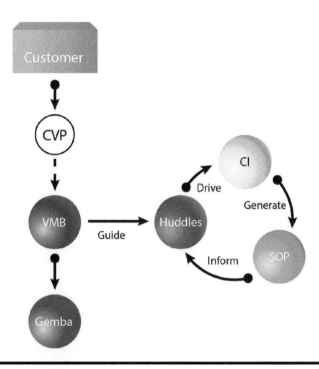

Figure 7.4.3 The CBA Habits Working as a System.[4]

for each direct report to the CEO and, more importantly, targets for the culture build.

It was mandatory for all general managers with teams greater than one hundred people to achieve the Bronze level within one year and they were expected to maintain this minimum level. The number of teams to reach Silver and Gold were not mandated but several group executives set their own targets for Silver accredited teams within each financial year. Also the concept of Platinum was introduced for teams that were at a Shingo prize level but did not want an external Shingo assessment due to various reasons.

Here are a few questions the sponsor should consider asking.

1. **Why would I go for Silver?** You will have seen the benefits for your business that focusing on going for Bronze achieved. The key benefits of deciding to aim for Silver level maturity are, firstly, setting new sights on a higher goal; and then getting the team focused and capable of solving problems.
2. **Why would I go for Gold?** You will have seen the benefits for your business that focusing on going for Silver achieved. The key benefits of deciding to aim for Gold level maturity are, firstly, setting new sights

on a higher goal. Sometimes after achieving a long-established goal momentum can reduce and setting a new goal can help to maintain focus and enthusiasm. Secondly, increasing the level of benefits by lifting the sights and trajectory to achieve Gold.

3. **What is the value of lifting my maturity to Gold level?** Consider the benefits achieved from lifting to Silver level—imagine that increased to include all staff on a constant basis. People readily commit to additional effort when they are self-motivated rather than being directed.
4. **What are the criteria for Gold accreditation?** Typically, Gold businesses will be General Manager (GM) with fifty or more staff and have been Silver accredited for twelve or more months, ideally two years.
5. **Can I go straight to Gold level accreditation without achieving Silver first?** In short, the answer is no; however, lifting maturity brings benefits in the short, medium, and long term. Aiming for Gold is a good plan, and recognition of achieving Silver level is a benefit on the path to achieving Gold.
6. **Who will support my business to lift our maturity to Gold level?** We will provide support to your business along the way. They will provide an initial workshop, ongoing maturity insight meetings (MIM), connections within the Group, Community of Practice, and network opportunities external to the Group.
7. **What is a MIM?** A maturity insight meeting is an optional meeting held between members of the business team and a certified assessor from Group Productivity. Consider them as similar to a (group) coaching session. That means it could take a variety of forms—often walking the business floor talking to various people, asking them questions to understand how they are progressing, and providing insights to the business on opportunities or approaches.
8. **What support or resources will I have access to, to help my business on this journey?** Group Productivity will provide opportunities for education, exposure, coaching, and networking.
9. **Where is the project plan to lift my maturity to Gold level?** As with lifting your maturity to Silver, lifting to Gold requires an even greater focus on changing culture, mindsets, and behaviors. This will require an approach unique to your business, based on your depth of understanding of the differing requirements of the various parts of your team.

10. **What is community contribution?** Part of building enterprise excellence requires leadership humility—an openness to share with and learn from everyone. This could include helping others with your knowledge, experience, and insights, as well as learning from other fresh eyes on your business. For example, through providing an "open house" where others can come to see what and how you have achieved your success or presenting at a Productivity Leadership Workshop to share your experience with other leaders. It could also be participating in a Community of Practice with others focused on building their maturity.
11. **Why do we include an external assessor for Gold-level accreditation?** We are focused on maintaining the integrity of our assessment process as well as re-affirming the credibility of our internal program with external recognition.
12. **What happens after I achieve Gold-level accreditation?** First and foremost, we will all celebrate your success! You will be one of the most mature business units with CBA and will be recognized for your achievement.

    You will continue to reap the rewards of the "way of life" maturity. You will be sought out as a leader in your field and you will have already helped to shape the program for the rest of the Group. At that level of maturity, improvement will be in the DNA of your business and thus maturity will continue to be enhanced.

    You may also choose to seek further external recognition from the Shingo Institute which will continue your maturity development. This is a significant decision (with considerable preparation and expense) and more details can be found on the Shingo website.
13. **Why are the Group's values important for lifting maturity?** An organization driven by values, with clarity on the behaviors, is evidence that it has truly reached Enterprise Excellence. Teams whose behaviors are driven by values are self-sustaining and self-improving, as constant customer focus, mature process management, and tenacious process improvement are simply a way of life.

What Were the Desired Behaviors Being Measured?

The leadership team understood that a key element of the productivity program success was going to be a change in employee behavior and this needed to be measured through the accreditation program.

The assessment system comprised high-level definitions of each dimension. The productivity team defined five key dimensions of culture assessment of behaviors: customer focus, production capability, process management, mindsets and behaviors, and product leadership. This is discussed in more detail in Chapter 8.

What Was the Design of the Assessment System?

To support this accreditation program, a quick reference guide was generated to help everyone develop and grow the maturity of each dimension.

Customer Focus

Customer Focus is about understanding what our customers value and consistently using that knowledge to make better decisions. What was trying to be assessed:

- There are documented processes throughout our business to regularly seek feedback on what our customers value, and how well we are delivering this.
- Each team in our business has developed a formal Customer Value Proposition (CVP), defining how they deliver value to the customer.
- In our business, we regularly measure and track metrics that impact customer satisfaction—for example, number of errors, complaints, response times, etc.
- In our business, each team regularly reviews the metrics that impact customer satisfaction to identify opportunities for improvement.
- We work with our internal business partners to help them understand what our team does and how we add value to our customers.

As a leader of a team, it is worth trying to answer:

- How does your CVP drive the behavior and thinking of your staff?
- How closely do the measures on the VMBs across your business align with your CVP?
- When did you last test your CVP against customer feedback?
- When was your CVP last discussed by leadership?

How do you know whether your processes are delivering against your CVP?

How do you understand the moments of truth that your customers experience?

To help build momentum

- Gather some VOC
- Test your CVP with customers
- Check how CVP measures align with the measures on VMBs
- Agree and schedule what other actions are required

Process Management

Process Management is about applying end-to-end process thinking and managing process performance using customer-centric metrics. What was trying to be assessed:

> In our business, we have developed process maps of our core processes that truly reflect the work completed within each process.
> In our business, SOPs are readily available to users.
> In our business, we regularly measure and report on the performance of our key processes using VMBs.
> Our VMBs adhere to the Group standard of three core components: performance measures, CI, and actions.
> We work closely with upstream and downstream teams that are part of the end-to-end process.
> In our business, the handover points of our processes to and from other departments are well defined and clear to everyone.
> Within our business, each core process has a process owner who regularly reviews process performance, identifies opportunities for improvement, approves changes to the process, etc.
> In our business, each core process has a process governance body led by the process owner, which brings together all the stakeholders that impact the performance of the process.
> Accountability is clear at all levels of our business.

As a leader of a team, it is worth trying to answer:

How are your processes measured? Averages or variation? End-to-end vs silo?

What efforts has your business undertaken to understand the relationship between leading and lagging metrics?

How clearly aligned are the metrics between various levels of VMBs in your business?

How is process thinking/documentation leveraged to add more value?

How often is the adherence to SOPs confirmed?

How do people know when to take action and when processes are performing?

How do people identify which process step is the next target for improvement?

What methods do you use to coach end-to-end process thinking?

How do you know you have optimized the way your processes are managed?

To help build momentum

- Centralize SOPs and agree on the amendment process
- Agree on level of process maps required
- Assess what needs to change on your VMB
- Agree and schedule what other actions are required

Productivity Capability

Productivity is about using productivity tools because they help us get better outcomes by adding more value, prioritizing effectively, and creating tangible benefits. What was trying to be assessed:

In our business, there is a well-defined Productivity program and everyone is encouraged to identify and get involved in opportunities for improvement.

In our business, we use a standard Productivity framework (e.g., PDCA, DMAIC) to solve problems.

Progress on Productivity initiatives/CI including the number implemented and the benefits delivered is highly visible and regularly communicated at all levels of our business.

In our business, we are encouraged to challenge existing ways of working.

As a leader of a team, it is worth trying to answer:

What are my expectations of the frequency of use of tools?
When did I last role-model using the tools?
How can I better utilize the capability already established in my business?
Are you focused on activities that add value to the customer (or any of the 4 Productivity questions)?
What is the root cause of that problem you are working to solve?
How can I help you to build your productivity capability?

To help build momentum

- Agree which key tools you want everyone to use for problem-solving
- Role model using the tools
- Expect, and regularly enquire about, the use of tools in problem-solving
- Agree and schedule what other actions are required

Mindsets and Behaviors

Process Management is about knowing what productivity means and understanding the benefits it has brought to our customers and our business. Consistently applying relevant tools to improve our business. What was trying to be assessed:

I believe Productivity is about making things simple and easy for customers and each other.
I know how my work contributes to achieving the CVP.
I understand how the metrics on our VMB measure our progress to achieving the CVP.
I actively use process maps, for example, to understand where I contribute to the process, or to help with CI.
I know which productivity tools I am expected to use to solve problems
I use productivity tools to solve problems.
I measure and record the benefits that my CI's produce.

As a leader of a team, it is worth trying to answer:

How many times have I visited each of the hurdles in my business?

How many of my staff have learned from my, or another team's, huddles?
How can I recognize and encourage more focused CI?
What does productivity mean to you?
What benefit did you get from the last CI implemented?
What do you think we need to do to improve the way we use the four productivity habits?

To help build momentum

- Agree and implement a method for recognition of CI
- Implement a method for measurement of CI
- Help everyone to have clarity on what productivity means in your business
- Agree and schedule what other actions are required

Productivity Leadership

Productivity leadership is about actively championing the value productivity will bring. Using inquiry-based coaching to drive better performance outcomes. What was trying to be assessed:

Managers and leaders in my business actively champion Productivity by going to where the work is done to understand what really happens.
Managers and leaders in my business role model Productivity by using the Productivity tools to solve problems.
Managers and leaders in our business use an inquiry-based coaching style where they help people understand and solve problems for themselves rather than telling them the answers.
Leaders in our business show that they value Productivity capability through encouraging development and utilizing their certified Green/Black belts.
Leaders in our business provide clear, consistent, and frequent communications about business direction and strategy.
The results from each accreditation survey are used to create new action plans.

As a leader of a team, it is worth trying to answer:

In what ways would my team say I role-model productivity?
How visible am I in championing productivity?

How are the messages I send about productivity cascaded through my business?
When did your leader last visit your huddle?
How often do you receive coaching for your own development?
What do you see happening across our business in relation to productivity?

To help build momentum

- Communicate clearly and frequently WHY you are lifting maturity
- Ensure coaching is provided to all
- Monitor your progress to plan
- Agree and schedule what other actions are required

The outcomes of Silver and Gold assessments were communicated directly to the management team who underwent assessment in the form of a written report that included:

strengths and opportunities identified during the maturity assessment;
a visual representation of progress against strategic targets for each area assessed; and
the consolidated answers to the Magic Wand (a catch-all) question aimed at defining what "must" be changed to better engage CBA people.

What Was Your Deployment Plan for the Assessment Program and for Developing the Assessment Program and Business Maturity?

The accreditation program was launched once the Bronze process was defined, with only a high-level concept of what Silver and Gold would look like. As lessons were learned going through the role out of Bronze, these could be built into the detailed definitions of Silver and, once Silver was starting to be rolled out, further detailed definitions for Gold could be determined.

Consistency with the messaging through the definitions of each of the phases was the key. Here is the communications summary used for all front-line staff.

Productivity continues to be a strategic focus for the Group. Being part of a team that is leading the way in Productivity in the Group, that has deeply embedded this capability and culture means:

- you receive education, experience, and exposure to important practices that can enhance your career development;
- you have the opportunity to obtain skills that are highly valued by the Group;
- you are improving the way you work for your customers, your colleagues, and yourself every day; and
- you are securing an efficient, effective, and productive business.

7.5 Bakkavor Desserts—Measuring the Maturity of the CI System Case Study

John Bowman and Leighton Williams

Using a system maturity assessment to build an understanding of what is required to develop an environment that will support sustainable operational excellence.

Context

In June 2018 Bakkavor Desserts at Highbridge embarked on a performance improvement initiative—the Go the Extra Mile (GEM) program. The intention of the GEM program was to develop an environment that supported a culture of ongoing improvement while delivering in excess of £1 million of in-year operational benefit.

The Organization and Type of Assessment

The client is a market leader in the provision of desserts to well-known high street supermarkets in the UK.

The site leadership team had been challenged to increase their throughput from £58 million to £75 million.
Project GEM was initiated to accelerate improvement in site performance. Specifically, to deliver £1.2 million in operational savings.

There had been previous CI initiatives over the years that had failed to deliver the promised benefits.

LWA consultants were challenged with facilitating the delivery of benefits in the following areas: labor efficiency (£550k), product disposals (£450k), and material usage (£200k).

To create the environment for the benefit delivery to be sustained and continually improved we had to create new ways of working and coach the leaders at all levels to adopt the behaviors that would be necessary for them to operate in this new environment.

The program was designed around the model in Figure 7.5.1.

We ensured that each aspect of this model was represented in the design of the program, which we adapted from the "Shingo Enterprise Excellence Model." Above all, for us to be able to sustain improvement we had to engage our people on the journey.

Once the program was designed, it was deployed by means of the method described in Figure 7.5.2.

We used this method to understand the current situation, describe where we wanted to go, and engage all departments, not just production, on the journey. Key leaders at all levels were encouraged to take the lead. Our teams and their leaders were challenged, coached, and supported. The program was a success, and the following was achieved:

The results were delivered ahead of schedule.

A management operating system (MOS) was embedded which facilitated ongoing improvement.

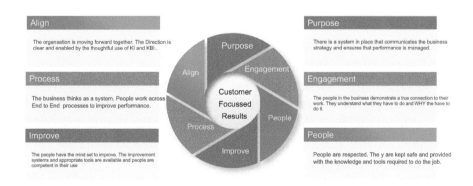

Figure 7.5.1 The Assessment Model.

200 ■ *Why Bother?*

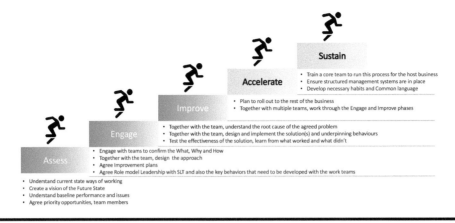

Figure 7.5.2 Program Design and Deployment.

Noticeable behavioral change was modelled around described and observable behavior.

Training aligned to develop future behaviors.

Operational teams learned about OPEx disciplines by means of A3 learning sessions and applied the learning through structured rapid improvement events (Kaizen Events).

Cross-functional teams worked together to improve NPI and planning processes.

Material Kanbans were established to control material usage.

On-the-job coaching around performance meetings and problem-solving (RCA) drove continual improvement.

Accreditation to Silver Level on the major high-street retailer Customer Scorecard.

Why the Assessment?

So, great results guys! Now the real challenge! How are we going to continue to develop as a business?

Yes, the performance KPIs would provide an indicator of the results, the outputs. What about the "soft" stuff, the activity that we KNEW made the results possible in the first place? We needed a mechanism that would be the antecedent for the Site Leadership Team (SLT) to focus on the inputs, the mindsets, behaviors, and routines.

As part of the initial value stream assessment, we carried out a high-level maturity assessment based. The result was not widely shared at this point. Based on an initial visit and discussion with the SLT we felt that the business was not yet mature enough in its understanding of Lean Thinking and Operational Excellence and that the output could be too much too soon.

During the program, we were assessed by the major high-street retailer Customer Scorecard and were awarded a Silver Level award for our efforts around Operational Excellence. This assessment process gave our SLT their first introduction to assessing systems maturity.

It was always the intention to assess the way the people at Bakkavor Desserts were working to embed the new practices. The business continued to improve and gained widespread acclaim within the group for its performance.

Around the beginning of 2020 we had some concerns that, while performance was still improving, some of the inputs, the antecedents, and routines were beginning to slip. It was felt that it was only a matter of time before this affected our performance. In addition, there was a strategic plan in place that would bring about significant opportunities for the business. We were concerned that the business could slip back to a "pre-GEM" position, which would impact our ability to deliver our strategy.

How the Measurement/Assessment System Works

Process and behavior both matter. With this in mind, we set out to measure each element by observing both aspects of the system:

The maturity of processes, procedures, and systems that had been introduced during the GEM program.
The maturity of the individual and collective behaviors developed during the program for which continuation had been agreed as part of the program handover and Phase 2 plan.

Our conclusions were reached by assessing what we observed against three criteria—extent, reach, and adherence—for the process and structure aspect of the system and another three—engagement, trust, and focus—for the behavior aspect (Figure 7.5.3). This would be applied to each of the assessment system sub-elements.

There are five levels of maturity ranging from Level 1 Innocence through to Level 5 Excellence. More about this later.

We applied this thinking to each of the system elements of Purpose, Process (and improvement), and People (leadership and engagement) outlined in Figure 7.5.4.

202 ■ Why Bother?

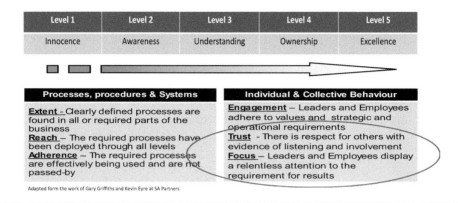

Figure 7.5.3 Maturity Levels.

We applied this thinking to each of the system elements outlined in Figure 7.5.4.

Purpose, Process (and improvement), People (leadership and engagement).

Each element had a fundamental baseline question.

Followed by three dimensions that defined the scope.

Then a statement that described what good looks like.

Figure 7.5.4 System elements.

Each of these three elements had a fundamental baseline question, followed by three dimensions that defined the scope, then finally a statement describing what good looks like.

Our conclusions were scored and presented as a summary (Figure 7.5.5). The target condition was highlighted in green and the current condition in pink. The target we took was the one that needed to be achieved for each element in order to achieve a Silver accreditation level with the major UK retail chain.

The Design of the Assessment—Our Approach and Principles

Why—this was made apparent to the senior leadership team, area leadership team, and selected line leaders.

The intervention was widely accepted and welcomed; however, there was some negativity from leaders who had not been involved in the original process. This quickly evaporated once the process began.

	Key question	Key Elements to Assess	Outcome Description of Element
Purpose	Is there a process in place to communicate the Strategy and align and manage performance?	Strategy and Business Planning	The vision, goals and strategy are clearly communicated throughout the organisation. Processes and departments are aligned to the stratgey by deployed key measures, targets and activities.
		KPI Deployment and Alignment	
		MOS management system	
Process Mgt and Continuous Improvement	Is the business using the approriate CI Tools and methods to improve?	Workplace organisation and visual management	The teams have a "daily habit" of continuous improvement and use simple, visual tools and techniques that have been chosen and adapted for their use
		Problem solving	
		Kanban and material management	
	Are we managing and driving Process Level Improvement acroos the business (GEM Projects)?	E2E opportunities for improvement are identified and managed	Opportunities for End to End Improvement is managed through processes and value streams in order to deliver sustainable imrovement and outstanding customer value with minimum bureaucracy and waste
		Equipment performnce and tthroughput	
		ETE material flow - MUV	
Leadership & Engagement	Are our People engaged in GEM / Continuous Improvement?	identifying ways in which their teams can improve	The team members feel that they can influence imrovement in their areas/departments. They are enthusiatic in planning and delivering improvement and openly share learning and celebrate the results of their activity.
		Rapid Improvement events	
	How well are we Leading GEM / CI at Highbridge?	Encourage learning through Delivering structured coaching	Our business Leaders (SLT and Band C) are activley involved in using structured techniques, coaching and conversations to develoip their people improve and deliver improvement
		Lead 'what matters' conversations	
		Actively promote / coach observable behaviour	
	MATURITY INDEX	MATURITY INDEX	Key Characteristics General description of characteristics that may be found at each of the milestones

Figure 7.5.4 System Elements.

The Principles of Our Approach

1. Being prepared to explain constantly and clearly the why, what, and how of the process.
2. Going and seeing for ourselves.
3. Conducting discussions at the Gemba.
4. Ensuring that we were appreciative of the work that had been carried out. What has been good over the period of the assessment.
5. Respecting people's opinions; seeking to understand.
6. Providing regular objective feedback and being hard on the process and soft on the people.

204 ■ *Why Bother?*

Key question		Key Elements to Assess	Description of Element	Level of Maturity					
					Innocence	Awareness	Understanding	Ownership	Excellence
				Score	0	20	40	60	80	100
Is there a process in place to communicate the Strategy and align and manage performance?	Purpose	Strategy and Business Planning	The vision, goals and strategy are clearly communicated throughout the organisation. Processes and departments are aligned to the strategy by deployed key measures, targets and activities.	Actual					
		KPI Deployment and Alignment							
		MOS management system		Target					
Is the business using the appropriate CI Tools and methods to improve?	Process Mgt and Continuous Improvement	Workplace organisation and visual management	The teams have a "daily habit" of continuous improvement and use simple, visual tools and techniques that have been chosen and adapted for their use	Actual					
		Problem solving							
		Kanban and material management		Target					
Are we managing and driving Process Level Improvement across the business (GEM Projects)?		E2E opportunities for improvement are identified and managed	Opportunities for End to End Improvement is managed through processes and value streams in order to deliver sustainable improvement and outstanding customer value with minimum bureaucracy and waste	Actual					
		Equipment performance and tthroughput							
		ETE material flow - MUV		Target					
Are our People engaged in GEM / Continuous Improvement?	Leadership & Engagement	Identifying ways in which their teams can improve	The team members feel that they can influence improvement in their areas/ departments. They are enthusiastic in planning and delivering improvement and openly share learning and celebrate the results of their activity.	Actual					
		Rapid Improvement events		Target					
How well are we Leading GEM / CI at Highbridge?		Encouraging learning through Delivering structured coaching	Our business Leaders (SLT and Band C) are actively involved in using structured techniques, coaching and conversations to develop their people improve and deliver improvement	Actual					
		Lead 'what matters' conversations							
		Actively promote / coach observable behaviour		Target					

Highbridge Bakkavor GEM (Opex / Lean) Business Model - Overall Maturity Index - August 2020

Figure 7.5.5   Assessment Scoring Summary.

The Process of Assessment

1. A remote briefing to explain the model and process. This had to be carried out remotely due to COVID-19 related restrictions.
2. SLT on-site briefing and initial group assessment.
3. Two days of observation in the factory (informal discussions as we went).
4. One-on-one structured discussion, either on-site or remotely.
5. Initial feedback session held in small groups around a key point summary.
6. Second feedback session on the detailed findings around each separate dimension. At this stage, it was important for consensus to be agreed on before moving to the action planning.
7. Action planning and next steps workshop.
8. Build outcomes into annual business planning/strategy deployment.

Delivering the Message

A potential issue with delivering the outcomes of an assessment is that it focuses only on the things that are not being done and does not structure any appreciation of the good work that has been achieved. It is absolutely important for acceptance and engagement that good work is recognized and seen to be appreciated.

We ensured the message was delivered to the SLT by means of the process in the previous section. What was critical for us was that the message be delivered in a way that stimulated discussion around the subjects. This could not be a TELL piece of work. To adopt that approach, however tempting, would fly in the face of what we had achieved in deploying ownership and accountability during the program.

Figures 7.5.6–8 provide examples of how the findings were presented to the SLT and key leaders.

Figure 7.5.6 represents the introduction to the feedback. A key point summary from the high-level indication of current condition vs target condition.

We used a SWOT (Strengths, Weaknesses, Opportunities, and Threats) format for each element. This allowed us to be appreciative of the progress that had been made—there was a lot!—while bringing focus to those areas

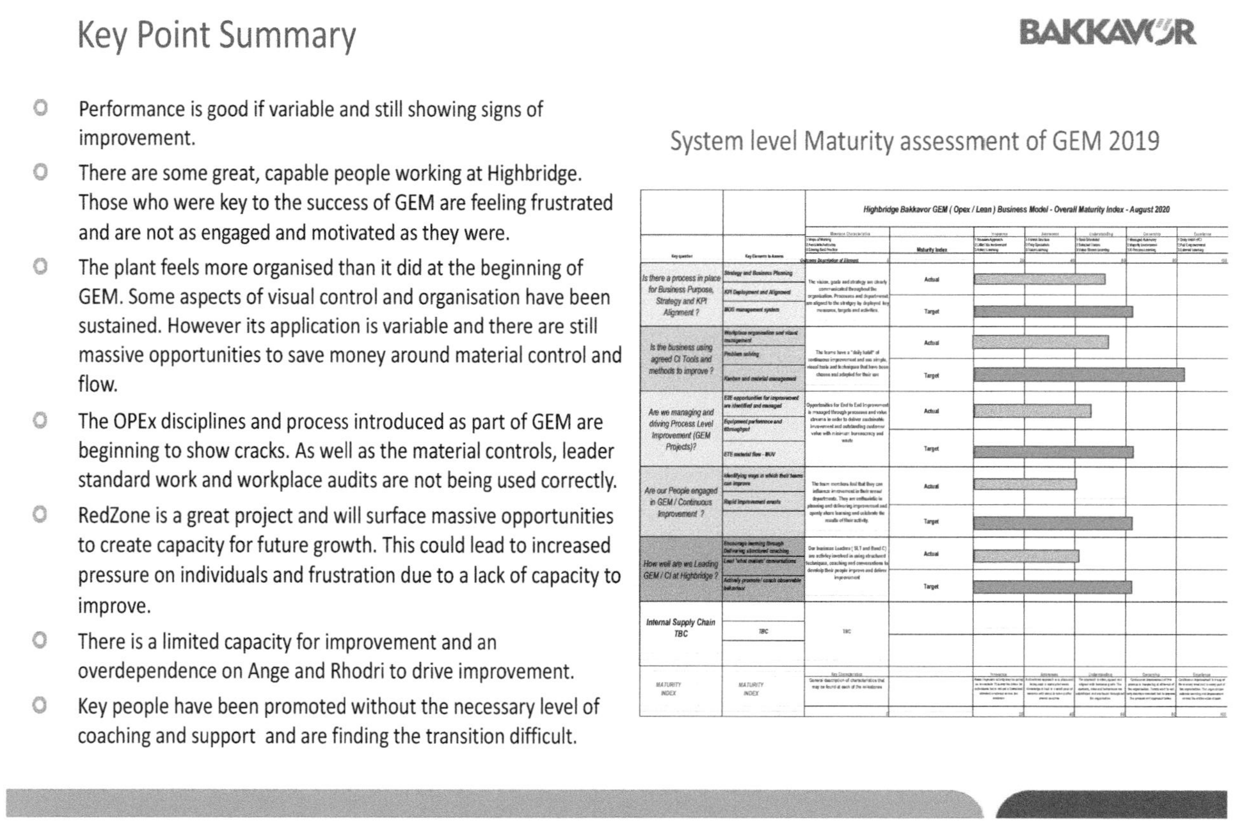
Key Point Summary
BAKKAVOR
Performance is good if variable and still showing signs of improvement.
There are some great, capable people working at Highbridge. Those who were key to the success of GEM are feeling frustrated and are not as engaged and motivated as they were.
The plant feels more organised than it did at the beginning of GEM. Some aspects of visual control and organisation have been sustained. However its application is variable and there are still massive opportunities to save money around material control and flow.
The OPEx disciplines and process introduced as part of GEM are beginning to show cracks. As well as the material controls, leader standard work and workplace audits are not being used correctly.
RedZone is a great project and will surface massive opportunities to create capacity for future growth. This could lead to increased pressure on individuals and frustration due to a lack of capacity to improve.
There is a limited capacity for improvement and an overdependence on Ange and Rhodri to drive improvement.
Key people have been promoted without the necessary level of coaching and support and are finding the transition difficult.
System level Maturity assessment of GEM 2019
Highbridge Bakkavor GEM (Opex / Lean) Business Model - Overall Maturity Index - August 2020
Maturity Index
Is there a process in place for Business Purpose, Strategy and KPI Alignment ?
Is the business using agreed CI Tools and methods to improve ?
Are we managing and driving Process Level Improvement (GEM Projects)?
Are our People engaged in GEM / Continuous Improvement ?
How well are we Leading GEM / CI at Highbridge ?
Internal Supply Chain TBC
TBC
TBC
MATURITY INDEX
MATURITY INDEX
Actual
Target
Actual
Target
Actual
Target
Actual
Target
Actual
Target

Figure 7.5.6 Introduction to the Feedback.

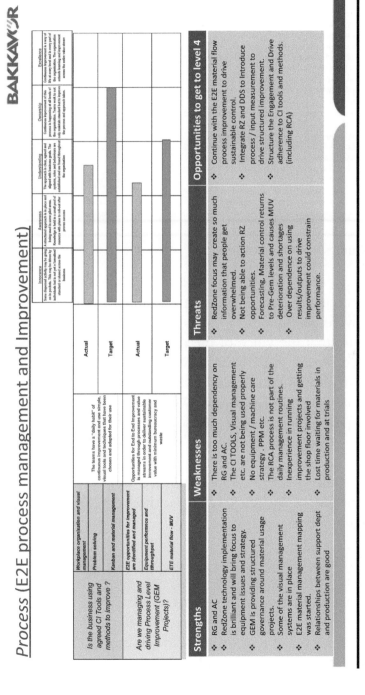

Figure 7.5.7 Example Findings.

208 ■ *Why Bother?*

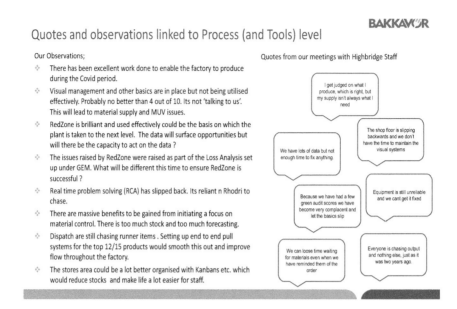

Figure 7.5.8 Quotes and Observations.

that needed attention. Figure 7.5.7 is an example of how the findings of each element were presented.

We made sure that the voice of the people working in the business was represented during the feedback. We shared quotes (Figure 7.5.8) that had been made during our discussions to provide context and insight into our conclusions and discussions. We asked the people who provided the quotes for permission to share these, anonymously in most cases. In some cases, the people we spoke with wanted to deliver the message themselves!

In addition to the system structure, we also gave feedback at the close of GEM 1 on the use of specific CI tools and on the behaviors that were agreed to. (Figure 7.5.9)

Next Steps

It is all too common for leadership teams to focus only on the results, the outputs of our efforts. Of course they are important and even essential for the business. It's our belief that teams who focus only on results can be inconsistent in their performance and often plain lucky.

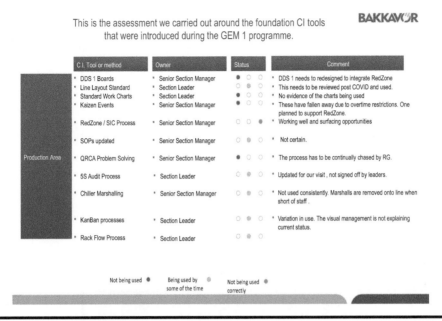

Figure 7.5.9 Status Tracker.

This activity reminded the team of the importance of working on the inputs and importantly, precisely what we have to work on, together that will ensure we build a platform for further improvement and sustainable success.

We have kept the GEM process going as a vehicle for focusing our improvement activity. We ensured that the key learnings from this exercise were integrated into our plan for 2021 GEM 4.

In some instances we need to revisit the basic disciplines, to go backward to move forward. In others, we build upon the good work and challenged ourselves to move onto the next level.

We set ourselves challenging targets and ensured that we were involved, supported, and coached the teams responsible for delivering the results in the work to develop the inputs.

During the final workshop, the team prioritized the areas needing further development. We facilitated this by connecting the opportunities that surfaced during the assessment (around Process, People, and Purpose) to the key elements of the business strategy as illustrated in Figure 7.5.10 and then produced the action plan illustrated in Figure 7.5.11.

210 ■ *Why Bother?*

	Issue / Opportunity	Impact on Strategy		
		Operational Excellence	People	Growing our business
Purpose	Celebrate SUCCESS AND Win Hearts and Minds	◁	◁	◁
	Reinvigorate GEM and align to strategy and loss meeting	◁	◁	◁
	Revisit strategy. Decide what to work on and what to leave	◁	◁	◁
	SLT to revisit their purpose (why) and decide how to work as one team (how).	◁	◁	◁
	Agree and Deploy observable behavior	◁	◁	◁
	Develop DDS 1/ perf mgt in support processes	◁		

Will addressing these issues have a positive impact on your strategy ?

Is there anything missing ?

Figure 7.5.10 Assessment Recommendations.

BAKKAVOR

Gaining momentum over the Next 10 weeks

Issue / Workstream	Owner	next steps	When
SLT back to factory floor and coaching (kata)	JB	• SLT reset • Decide how to coach 'band c' at succession review	• 15th Oct • 16 the
Refresh key CI tools and then take to the next level	RG	• Understand what tools are the top 5 and create a plan to take forward (the tools may include 5S, RCA, LSW & process confirmation)	• 15th
Align DDS 1 and DDS 2 and KPI alignment top to bottom	AC (AW)	• Reestablish the DDS1 process in production • Functional leads to sense check their own DDS 1	• 10th
Re-energise the E2E material analysis work and take it to the next level .	AW ?	• • •	• • •
Re-energise the MUV process and take to the next level	CB	• get the MUV meeting back up and running • educate night shift and establish monthly muv • material pips	• 10th • Nov 8 • Nov 8
Kaizen events to engage / celebrate success	PR	• A Plan of events for the next 10 weeks	• 15th
Communication (and engagement strategy)	JD	• develop what will work (discuss ideas with sel and mgrs.) and sign off at SLT •	• 23rd

Figure 7.5.11 Summary Action Plan.

Lessons Learned

From a business perspective, we realized that we had let our focus on leading proactive improvement slip. It is all too easy in an FMCG environment to revert to only chasing outputs and to lose sight of our need to promote the activities that will help us build the environment we need to facilitate ongoing improvement.

We recognized that as Leaders, we need the same (if not more) prompting and discipline around developing their behavior as their teams. We need to routinely assess where we are against our own commitments. Using this framework to make us take time out of the day to day to work on the system and develop our behavior is essential.

From an assessment perspective, we now understand the importance of seeking the view from our colleagues at all levels of the business. If you are considering using a maturity tool as we have, ensure all levels are involved. It takes time to ensure that your people are connected with WHY you are carrying out the exercise and requires that you ensure those that take part receive feedback on the output.

It requires time to carry out the appropriate level of preparation. It's worth it.

During the assessment take the time to really understand what you are seeing and being told. Listen attentively, revisit situations and discussions if necessary, to develop the understanding. Above all be flexible in your approach.

Don't forget the purpose of the exercise is to learn. Without learning there is no improvement.

7.6 Virtual Assessments Case Study

Morgan Jones

Introduction

An organization had a maturity assessment program that conducted assessments across both sites and functional areas. They conducted a baseline assessment at the beginning of their Lean transformation deployment at each site and function, and then subsequently every six months during the two-year deployment and annually thereafter in the sustain phase post-deployment.

The assessment had several steps:

1. Preparation, including:
    (a) communication with site management teams;
    (b) agenda setting; and
    (c) pre-read of site plans and operational and safety performance.
2. Site visit/assessment including interviews at all levels, observations, and on-site generation of a report.
3. Report out to the leadership team.
4. Action planning workshop.

The site visits usually consist of four to eight assessors physically on the site/function to conduct interviews, observe routines and meetings, assist with problem-solving and CI sessions, and conduct Gemba walks.

The Impact of COVID-19

When COVID-19 struck globally in 2020, the company implemented an action plan to manage business risk, including stopping all non-critical travel to sites and across state borders. This meant that the assessors could not travel to the site and the assessment teams were challenged to come up with an alternative assessment approach—they created virtual assessments. The key was to implement an approach using off-the-shelf, wearable technology so that on-site team members, or "avatars," could livestream video to the assessor group located remotely.

The Solution—Wearable Technology

The wearable technology selected had to operate both in the office and in harsh, remote mining sites. The team selected RealWear HMTs (head-mounted tablets). These could be worn attached to either a hard safety hat for operations environments, bump caps for maintenance workshops, or a simple strap/baseball cap for office environments.

The RealWear HMTs were selected as the wearable technology to enable virtual assessments because:

- The RealWear HMT-1 was a hands-free, voice- or gesture-controlled Android™ wearable computer for industrial workers. It was compatible with the company's safety and protective equipment policies and suitable for use in wet, dusty, hot, dangerous, and loud industrial environments. It provides the foundation for a robust and connected remote-worker program within the BOS.
- It was a fully-rugged, head-mounted device and had the option of snapping onto safety helmets or attaching to bump caps. It can also be used with safety glasses or corrective eyewear.
- The high-resolution micro display fits just below the wearer's line of sight and views like a seven-inch tablet. It shows the wearer either the video they are streaming, or saving video and audio to storage on the headset.
- It had GPS and Wi-Fi enabled and connects off-site assessors with on-site key stakeholders through motion stabilized, livestreaming video of OEI routines, voice contact via microphone and bone conduction speakers, and text-based chat via the boom-mounted display, all with no impact on other people on site.

Using wearable technology that was suited to the operating environment was important. The team determined that using a GoPro with a helmet mount may appear threatening and not provide the ability to liaise with our on-site assessors. We also worked through company regulations regarding the recording and use of video images, as people may be resistant to the capture of video. Hence, before deployment, a strong focus on the change management aspects of using the virtual technology was needed to allow on-site teams to understand how the technology will be used.

The other factor for successfully using technology in remote operations sites was reliability. The HMT is a stable technology. One challenge was the stability of Wi-Fi, and livestreaming in early virtual assessments became an issue. Consequently, we developed backup sources using hot-spotting of mobile phones and the ability to record all video and sound to memory on the HMT.

There are obvious elements of using technology that required some consideration:

- Ensuring headsets are fully charged and ready for use at the beginning of each session.
- Ensuring Wi-Fi availability.
- Training headset wearers to use the device (i.e., some simple voice controls for HMT).

HMT was originally designed for one subject matter expert communicating directly and only to the headset wearer. As such, the headset had noise-cancelling capabilities to maximize the clarity of the communications. However, it was eventually decided that it would be useful for the HMT to also detect ambient sound so that conversations could be heard in routine meetings, requiring the addition of an external Omni-directional microphone. This was blue-toothed to the HMT to allow video and sound to be livestreamed and/or recorded for the off-site assessors. The device selected was a Jabra 710 Conference speaker, which provided good quality sound that synchronized with the video stream.

Lessons Learned

Preparation and Pre-reads

Prestart preparation for an assessment was extended from six to eight weeks. This extra time was needed to manage the additional items required when using wearable technology on a mine site. These additional items were:

- Undertaking a formal risk assessment—in some jurisdictions this was a legal requirement.
- Identifying and training headset wearers to use the technology.
- Managing the logistics of getting technology to sites (aim for a minimum one week prior).
- Communicating and engaging with staff and unions regarding livestreaming and/or capturing of video and images.
- Introducing the front-line teams to the headsets in the prestart by the on-site team—a week prior—and being available to answer any questions.
- Planning time for remote assessors to watch any captured videos; Wi-Fi issues can lead to hours of recording having to be reviewed.
- Engaging a site contact to ensure the smooth running of assessments—enabling WebEx, chasing attendees, etc.

Site Visit Interviews with Managers and Front-line Staff

This required relatively little change. Instead of interviewing people in person, they used video conferencing facilities. There are many potential solutions for video conferencing; however, Cisco WebEx was the standard package available across all operating sites. Key lessons have been:

- Making sure to add a WebEx/Zoom meeting facility to electronic invitations.
- Having someone on-site to coordinate and remind people of the WebEx meeting, as they are mostly operational staff and may miss electronic invitations.
- Focusing on building rapport remotely, ensuring people feel safe to share. It was essential to reinforce both anonymity and that all answers are correct, they cannot be wrong.
- Reinforcing that the interview was NOT being recorded, just being streamed live to an off-site assessor, Unless there are wi-fi technical issues, then it may be recorded.
- Identifying clearly who was the lead assessor for interviews, with clear signals/handovers so that additional assessors can add questions where applicable.
- Allowing additional time between interviews allows the assessors to gather their thoughts and write notes.

Site Visit Observations of Routines and Meetings

Observing the team's behaviors in action and confirming verbal explanations from interviews was an essential element and source of evidence for assessments. Not being present in person to conduct interviews raised a few challenges to assessments, so the use of the wearable technology that allows remote assessors to see and hear was essential. The key learnings in the use of wearable technology for observing routines and meetings, undertaking pre-starts, and shift change have been:

- Ensuring that the site has reliable Wi-Fi connectivity.
- Having at least three levels of redundancy when it comes to capturing routines:
  - Wi-Fi
  - hot-spotting to phone
  - recording directly to the headset
- Ensuring that you have completed the risk assessment and follow any controls/conditions required.
- Implementing a COVID-19 cleaning procedure that complies with OEM guidelines.
- Ensuring that the headset wearer:
  - has enough time to comply with the agenda (allow travel and set-up time)
  - arrives early to the meeting to check the Wi-Fi connectivity
  - locates themselves in a good position to adequately capture video imaging
  - places the microphone in an ideal position to pick up the team conversation
  - moves their head smoothly and slowly to improve video vision capture

Site Visit "Go and Sees" (Gemba's)

As with observing team meetings and routines, the ability to "go and see" or do Gemba walks around the areas where work was undertaken was another source of evidence and was an integral part of the assessment process. It was important to predetermine both the physical location of Gemba walks and what you are looking for to guarantee that you have a Gemba with Purpose. Key lessons have been:

- Having the onsite contact take photos of visual boards and points of interest to add to observations, guided by the off-site assessor.
- Providing additional on-site resources if there are connectivity issues, guided by the on-site assessor on what to view and/or questions to ask people on Gemba.

What Did not Work?

- Reading body language during interviews over WebEx and livestreaming via HMTs.
- Some interview techniques, such as non-verbal cues, elicit more information.
- Room set up, or even conducting interviews in the interviewee's environment to put them at ease, was not always available.
- After observing routines/meetings, not then being able to walk past visual boards using HMTs to closely observe recent and up-to-date information. In person, it was easy to talk with individuals and deep dive into what the information on the boards means to them.
- Trying to deep dive into evidence of previous CI/RCAs.
- Applying the 1:3:10 rule around visual boards. Within **1** second know if the team was winning or losing, within **3** seconds which part of the team needs focusing on and **10** seconds what actions and improvements are in place to close the gap.
- Having inexperienced headset wearers not understanding what was to be observed.
- Agenda and interactions being "staged," no ability to impromptu chat with a front-line worker.

Lessons Learned Summary

1. Using technology:
   - Check any corporate policies and jurisdiction regulations on permissions and capturing video.
   - Check for Wi-Fi connectivity to enable live video-streaming.
   - Ensure large SD cards to record large amounts of videos as a backup.
   - Develop a COVID-19 hardware cleaning process, as there are multiple users.
   - Consider conducting a risk assessment.

- Have a change management and communications plan for introducing wearable technology usage.
- Test technology on multiple occasions prior to assessment, including both the on-site wearers and the off-site assessors. This should include wearer access, training in the use of the equipment, and internet access.
- Ensure that the headsets are charged overnight, before, and during the assessment.
- Ensure that the chosen technology provides:
  - high-resolution video
  - high-quality audio

2. Changes in planning assessments:
   - Allow more time for logistics planning.
   - Plan to have technology available early at the assessment location for testing and local training.
   - Plan to capture images of visual boards to check they are up to date.
   - Provide a preview of technology to teams prior to the assessment commencing.
   - Allow extra time in each interview to build rapport with the interviewee and put them at ease with being on camera.
   - Ensure a daily check-in and check-out with the on-site participants to build alignment and get on top of any emerging issues.
   - Allow extra time in the agenda for the on-site team to upload video and photos for the remote assessors.
   - Build-in some additional daily check-in time so the assessors can discuss what they are seeing so far and where the gaps might be that need filling the next day. This needs to be more deliberate when the assessment was being conducted remotely.
   - Try to include a "surprise interview/meeting" to ensure unplanned engagement.

3. Changes to observing behaviors in a virtual environment:
   - Stronger focus on listening skills.
   - Danger of seeing only what the site wants you to see.
   - Focus on building rapport virtually to create a safe environment.
   - Ensure understanding of who was in the room when you observe. Most people shy away from the camera and thus tend to stand behind the wearer.

- Explain that questions will be asked and that these will be posted either directly to the group or via the wearer.
- Encourage the wearer to ask relevant questions and to get in close to the visual boards so that you can see the details; remember that this takes additional time.
- Consider the possibility that some areas that may have been identified when on-site may be hidden when off-site.
- Still allow time for flexibility in who to interview—for example, attend a virtual meeting and then pick people to interview who were at the meeting.
- Allows assessors to spread out the interviews over several days.
- Pre-recording of meetings also works well so you can watch these and then select additional interviewees.
- Opportunity to use the headsets to do virtual "Ohno circles," where assessors just watch an area.
- Challenge evidence and scores for behaviors/systems so as not to overinflate or penalize observation scores.

4. Giving feedback to leadership teams on results and observations:
   - Take extra time to set context clearly; what you will share, what feedback/questions you will seek from them.
   - Provide feedback in blocks with plenty of time for the leadership team's input and questions.
   - With limited opportunity to read body language, listening closely to responses and questions helps to better understand the true reception to the feedback.
   - There was the potential to show videos from the assessments to demonstrate examples of points being made in the feedback.

Chapter 7 Overall Key Takeaways

1. There is no one, perfect approach that will work universally in all organizations. One size does not fit all.
2. There are lots of opportunities to learn from what other organizations have undertaken and built on their experiences.
3. Be prepared to constantly learn and adapt your approach to designing, using, and improving your CI assessment system.

Notes

1 Image source: Forbes.com (2015). Panalpina commits to continuous improvement.
2 Developed by SA Partners. www.sapartners.com.
3 Jones, Butterworth, and Harder *4+1 Embedding a Culture of Continuous Improvement* 2nd Edition (Action New Thinking, 2018).
4 Jones, Butterworth, and Harder, *4+1*, 37.

Chapter 8

Why Bother Aligning Assessments, Assessors, and Calibration?

Chapter Summary

It is important to have consistency in the assessments to show they are both robust and credible to ensure the recommendations will drive the best focus for action planning. While there is always going to be a certain level of subjectivity it is important to make the assessments as consistent as possible to avoid potential confusion and mixed messages. Each assessor must be trained in-depth to a standard that can be universally applied but at the same time recognizes their expertise and experience. The customer experience must be consistent.

It is important to understand that one of the foundation stones of assessments is the credibility of the assessment process and the subsequent result. This requires impartiality, transparency, and humility on the part of both the assessment team delivering the assessment and report and the leadership team receiving the assessment and report. Establishing the integrity of the process, the assessors, and the calibration of the assessment results is one of the first things required to begin the assessments.

8.1 Why Consider the Maturity of the Assessment?

The purpose of the assessment is to measure the maturity of current business behaviors during and after a transformation, including when implementing Lean. It is also important to evolve the assessment program as the deployment matures. Figure 8.1 is an example of the pathway to maturing the assessment program. It shows how the assessment program is perceived through the various stages of maturity—for example, from pure compliance through to application, site, and finally enterprise embedment.

Having a maturity roadmap helps the team and organization start the assessment process. Rather than being a purist, start with mapping out where the organization currently is and then, as the maturity of the organization increases, the maturity assessment program can also mature. That way, people do not become disheartened due to an excessively low score and can realize that it is a measure of the management routines and practices rather than business outcomes—that is, not just WHAT is delivered but HOW it is being delivered.

8.2 Why Have a Process for Assessments?

It is important to have a formal process for assessments. There are three major phases:

1. Assessment Approval and Preparation.
2. Running the Assessment.
3. Assessment follow-up.

Figure 8.2 shows a ten-step process for setting up and running maturity assessments.

8.2.1 Phase 1—Assessment Approval and Preparation

The outcome of this phase is to have a fully defined schedule of interviews, meetings, and infield interactions (Gembas), accompanied by all pre-read materials as per the agreed deployment goals.

Assessment Maturity Pathway

Guiding Principles:
- Breadth across value chain
- Depth in Organisation
- Frequency of interactions
- Intensity of engagement with Maturity Assessments

Level 1 = within own team/site
Level 2 = some comparison between sites
Level 3 = comparison within business unit/site and some across enterprise
Level 4 = comparison across the enterprise, but within industry
Level 5 = External comparison industry agnostic

	Compliance Level 1	Explore Level 2	Application Level 3	Site Ownership Level 4	Enterprise Embed Level 5
Intent	Maturity Assessments viewed as an action to be completed; more of an audit used by the management team	Maturity Assessments are completed in adherence to a program schedule and are seen by some teams as an opportunity to improve.	Maturity Assessments are requested by some teams to accelerate the path to achieving their strategy.	Maturity Assessments is seen as a critical path to learning and growth at the site/department level. Focus on local results	Maturity Assessments is seen as a critical path to learning and growth by all levels of the business and maintains focus on business results and future aspirations
Output	Focus on score and top-down, one-way dialogue; leaders across business areas over-focus on score to the extent of "ignoring" value feedback for improvement	Focus moves to the Maturity Assessments Action Plan and opportunities to improves core processes.	Maturity Assessments Score and feedback seen as equally important	Feedback is seen as important output of Maturity Assessments and focuses on continuing to raise the bar". Differences across teams used by leaders as a way to learn from each area	Feedback is seen as the most important output and focuses on continuing to raise the bar "and what the aspirational benchmark might be". Differences among business areas used by leaders as a way to learn from each other
Follow-up	Unstructured, dependent of local site leaders and coaches; central team follows up periodically to find out what actions have been taken	Structured and driven by local management; starts to focus on longer term outcomes	Driven by local self managed teams who engage with senior leaders to ensure enterprise alignment	Site LT integrate Maturity Assessments action plans into their planning cycle (at appropriate level); central team viewed as a business partner o facilitate improvement	Business areas and leaders integrate Maturity Assessments action plans into their planning cycle (at appropriate level); central team viewed as a partner who looks externally to help build next level of improvement
Business Partnership	Is a push by central team to the business	Local management teams establish beneficial internal partnerships, and start to consider assessments on a shorter cycle	Being built into management routines and input to strategic and operational planning. Some teams developing specific KBIs	A pull from site LT and viewed as an integral part of site performance. Valued as a strategic KBI Some enterprise integrations	Is a pull by the business; viewed as an integral part of business performance. Valued as a strategic KBI
Internal Assessors	Recruiting based on who is available; assessment checking what is or is not there; more one-way communications/readout	Assessors are identified as key roles and valued as developmental opportunities.	Business leaders view Maturity Assessments as a strategic and essential "check" in their high level PDCA. Assessments are embraced and welcomed with people keen to undertake assessments outside their teams and sites for the learning experience and receive feedback.	Assessor role seen as a future leaders traits/skills Assessments are proactively requested and welcomed with people passionate to undertake assessments and learn.	Recruit high potential future leaders, using the key traits/skills as initial level screen to enable to above
Engagement Level	Centre Team High Site LT : Medium Mgr. Low Front line None	Centre Team High Site LT : High Mgr. : Medium Front line: None	Centre Team Medium/High Site LT : High Mgr. Medium Front line Low	Centre Team Medium Site LT : High Mgr. High Front line Medium	Centre Team Low/Medium Site LT : High Mgr. High Front line High

Figure 8.1 An Example Assessment Maturity Roadmap.

224 ■ *Why Bother?*

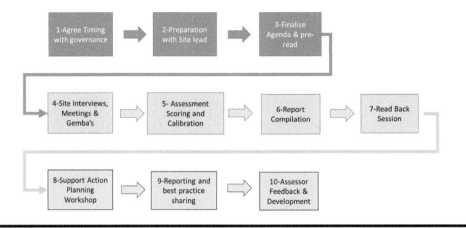

Figure 8.2 Example Maturity Assessment Process.

8.2.1.1 Agree Timing with Governance

Some form of governance with major milestones and tasks regarding time and cost should be monitored in the organization, this usually resides in a Project Management Office (PMO).

8.2.1.2 Preparation with Site Lead

The assessment coordinator will work with the site lead (or delegate team) during the thirteen weeks prior to the planned assessment.

8.2.1.3 Finalize Agenda and Pre-read Information

The assessment agenda is built by the business team. The site lead (or delegate team) will work with the lead assessor to ensure that the agenda and required pre-read materials are available to the assessment team within the required time frame.

8.2.2 Phase 2—Running the Assessment

The outcome of this phase is to gather data and evidence to be able to provide recommendations to the leadership teams on the maturity score, strengths, and opportunities.

8.2.2.1 Interviews, Meetings, and Gembas

Gathering information and evidence of the current business operating routines and habits is the core purpose of the interviews and direct observations (meetings and Gembas) during the assessments.

People from all levels of the function/site are spoken to, from the general manager through to the leaders, manager, supervisors, and front-line workers.

There should be a fully populated schedule completed prior to the site visit. This requires having people booked into meetings, with WebEx meeting invites attached, and the recipients accepting the session.

It is important to have a flexible schedule during the site visit, as interviewees' availability may change. The site team will have to coordinate the attendees for interviews, as some front-line staff does not have access to emails. In addition, an important part of the assessment is that the site team will need to provide escorts for the visiting assessors and transport them to the various locations.

During the Gembas it is important for the assessment team to confirm the focus of the Gembas so that they can conduct 'Gemba with Purpose'—that is, observe work areas, 5S, interview front-line workers in their workplaces, and focus on specific evidence for specific behaviors and/or practices.

It is important that participants have an open and frank conversation with the assessors during interviews, giving examples of various behaviors. It is important to use the meeting observations and Gembas to see the behaviors in action to corroborate the described behaviors in the interviews.

8.2.2.2 Assessment Scoring and Calibration

The scoring process is the result of all the notes, data, evidence, and observations collected through interviews, meetings, and Gembas.

Scoring starts at zero or one, where there is no evidence, through to the highest level (four, five, or ten), where the maturity is world-class. For each of the desired behaviors, use the framework definitions for each level of maturity. Each assessor will score the desired behaviors individually. Half scores can be given when some of the next maturity behaviors have been observed but not enough to lift it fully into that next maturity level. If the assessor does not have enough evidence to score any question, then it is appropriate to select "no observation" within the tool, resulting in no score for that question/behavior.

8.2.2.3 Report Compilation

The report presented to the site leadership team will display the overall maturity level (potentially in the form of a score). The wording of the strengths and opportunities should be action-orientated, challenge thinking, and start with an action word.

Ensure that there are documented strengths associated with the highest dimension of the desired behaviors.

8.2.2.4 Read Back Session

The purpose of the readback session to the site leadership team is to give key strengths and opportunities for each dimension of the desired behaviors. The readback report is used to develop an action plan to prioritize the focus of improvement to the site operating system and behaviors; however, the opportunities can be quite specific. An example could be to roll out front-line improvement visual boards to enable front-line staff to identify, prioritize, and track the progress of the ideas that they can influence within their control.

The key is to provide the site leadership team with opportunities to provide specific guidance in the report and verbal examples from the assessors to help guide their action planning.

8.2.3 Phase 3—Assessment Follow-up

The outcome of this phase is to update best-practice sharing across deployments and to continue to develop the central and part-time regional assessors.

8.2.3.1 Support Action Planning Workshop

A key deliverable to the business is to transform the assessment report into tangible leadership team actions. The action planning workshop is where the leadership team reviews the strengths and opportunities from the assessment report, brainstorms potential options, aligns other plans and actions, and assigns ownership and timelines for the actions to be delivered.

There will be opportunities for the dimensions in the report to be addressed, as well as a summary of the "magic wand" questions to be considered. It is not necessary to have actions against every opportunity, but rather that the leadership team should prioritize the easiest/biggest impact actions they want to work on first.

It is recommended that the leadership team makes the assessment report available to the site/function and shares the action planning in line with the desired behaviors to promote transparency and accountability.

8.2.3.2 Reporting and Best Practice Sharing

There are two key messages to be communicated post-assessment:

1. To provide a summary of each assessment to the broader BOS teams.
2. The best practice sharing gives a reference point for all existing and developing BOS deployments to review where the top-performing sites/functions are to view and discuss by each dimension.

8.2.3.3 Assessor Feedback and Development

At the beginning of each assessment, the lead will agree with each assessor on the focus areas for experienced and onboarding trainee assessors. During the assessment calibration and report feedback, the lead assessor will gather feedback for strengths and improvement opportunities from all of the assessors.

Each assessor will also complete a self-assessment of their assessor skills that will be validated by the lead assessor. The results of the assessment calibration and skills assessment will be referenced to the individual assessor.

8.3 What Are the Suggested Assessment Team Roles?

Lead Assessor—The person accountable for the progress of the assessment program maturity and coaching assessor development. They are accountable for setting up and improving the maturity assessment grid. They also develop and improve the assessment standard work/process.

Assessment Lead—The person who heads an assessment team at a site or facility.

Assessors—The assessment team members who gather material from interviews, observations, and Gembas to compile sufficient information to assess the behavioral maturity according to the standard.

8.3.1 Lead Assessor

Lead assessors lead the assessment team development and overall program in that region. They will develop, maintain, and mature the team's assessment program and assessment standard work and process.

8.3.2 Assessment Lead

The assessment lead will lead an assessment team through an assessment, engage with a deployment area, and determine the current maturity of the business.

Preparation with site lead:

- Confirm that the assessment coordinator is completing milestones and assist with feedback when required.

Finalize agenda and pre-read:

- Work with the assessment coordinator to ensure that draft and final agenda reviews are completed.
- Attend a lead assessment team planning session prior to the assessment being conducted.
- Ensure the assessment team have all required Personal Protective Equipment (PPE), whether their own or supplied by the site teams, and has completed all mandatory training for any site visits.

Site interviews, meetings, and Gemba:

- Lead and mentor the assessment team through the assessment.
- Meet daily with the team to confirm observations and indicate areas to address in subsequent sessions.
- Responsibilities to address safety and respectful behavior issues remain during the conduct of an assessment.

Assessment scoring and calibration:

- Complete scoring of the ideal behaviors identified using the assessment grid that was developed as part of the standard work.
- Lead the assessment score calibration activity.

Report compilation:

Lead the preparation of the assessment report back presentation and create a summary of the strengths and opportunities and appropriate behavior observations.

Read back session:

Lead the assessment read-out presentation to both the site leadership team and the site team.

Support action planning workshop:

When requested, support the deployment through an assessment action plan workshop by providing insight into the highlighted strengths and opportunities.

Reporting and best practice sharing:

Provide insight to the site areas of best practice regarding opportunities observed within the assessments.

Assessor feedback and development:

Direct feedback on what the site did well and the areas for further improvement.
Update the assessors' skills self-assessment and coaching log to further develop the assessors.

8.3.3 Permanent and Embedded Assessor

The permanent and embedded assessors work within an assessment to engage with an assessed area and determine the current maturity of the desired behaviors.

Finalize agenda and pre-read:

Complete and submit their assessor's skills self-assessment to the lead assessor.

Work with the assessment coordinator to confirm that the draft and final agenda reviews are completed.
Attend any assessment team planning session prior to the assessment being conducted (including training sessions on virtual assessment technology).
Ensure they have all required PPE for any site visits, whether their own or supplied by the site teams.
Complete all training and visa requirements for site access.

Site interviews, meetings, and Gemba:

Actively participate in all assessment observations and interviews.
Participate in the assessment team discussions regarding assessment progress as the assessment is being conducted.
Offer support, where required, to facilitate the virtual assessment team's ability to participate in the assessment.
Comply with all site policies.

Assessment scoring and calibration:

Complete scoring assessment.
Participate in the discussion during the assessment score calibration activity.

Report compilation:

Participate in the discussion during the preparation of the assessment report presented back to the leadership team.

Read back session:

Attend the assessment report to the site leadership team.

Assessor feedback and development:

Complete an assessor self-assessment post-assessment.
Work on the assessor development and coaching log.

8.3.4 Guest/Trainee Assessor

Assessment observers are organization leaders that are interested in observing a maturity assessment to enhance their ability to lead the business in implementing BOS practices. This observer experience helps them better understand the assessment process without the requirements of being an assessor and conducting behavioral assessments. One observer will be allowed to join each assessment. The observer will always be accompanied by the lead or another certified assessor.

Preparation with site lead:

Actively participate in the lead assessor training session (trainee assessors).

Finalize agenda and pre-read:

Complete and submit their assessor's skills self-assessment to the lead assessor.
Confirm access to the assessment tool the through assessment coordinator (trainee assessors).
Attend any assessment team planning session prior to the assessment being conducted (including training sessions on virtual technology). Ensure that they have all required PPE for any site visits, whether their own or supplied by the site teams.
Complete all learning requirements prior to the site visit.

Site interviews, meetings, and Gemba:

Actively listen and observe during certified assessor-led observations.
Participate in assessment team discussions regarding progress of the assessment.
Offer support, where required, to facilitate the virtual assessment team's ability to participate in the assessment.
Comply with all site policies.

Assessment scoring and calibration:

Participate in the discussion during the assessment score calibration activity.

Report compilation:

Participate in the discussion during the preparation of the assessment report presentation.

Read-back session:

Attend the assessment report presented to the site lead team (trainee assessor—compulsory, guest assessor—where available).

Assessor feedback and development:

Complete an assessor self-assessment post-assessment (trainee assessor).

8.3.5 Assessment Coordinator

The assessment coordinator leads the planning phase of the assessment and works with the assessment team while the assessment is being conducted and post-assessment.

PMO agreed on deployment goals:

Confirm all assessment milestones are met and reported back to the governance team/lead assessor.

Preparation with site lead:

Work with the permanent and embedded assessors to confirm availability and eligibility to complete upcoming assessments.
Have regular contact with the site teams to confirm alignment with the forward planning assessment coming up, to keep track of any potential assessment time frame updates.

Finalize agenda and pre-read:

Discuss the draft agenda with the assessor team to confirm understanding and that achievement of a quality assessment can be obtained through all scheduled interviews, meetings, and Gemba.

Confirm that all pre-read materials and the final agenda are available to the assessment team.
Complete training sessions on virtual assessment technology, if required.

Site interviews, meetings, and Gemba:

Be available to assist the team where required while the on-site assessment is being conducted.

Support action planning workshop:

Assist with any planning requirements to enable the support action planning workshop.

Reporting and best practice sharing:

Report information to the wider improvement community about assessment results, including highlighting best practices.

8.4 Why Be Consistent Across Assessments?

Completing an assessment gives the organization/site a snapshot of the maturity of its routines and practices. Credibility of the assessment program is a foundation of the business uptake of assessments and their outcomes. It ensures that a consistent process is applied across different business areas, functions, and sites. There are two main elements to ensure this happens:

1. Defining the behaviors you are trying to measure.
2. Defining a measurement system—how behaviors will be scored consistently.

8.4.1 Defining Behaviors to be Assessed

As mentioned earlier in this book, there is a need to define the ideal behaviors you want. Once they have been defined then deciding how you can observe these behaviors consistently becomes a challenge, especially when you try to observe the behaviors at various levels of seniority and the behaviors start to change.

From our experience, it is important to spend time identifying the different behaviors at these various levels of the organization around "continuous improvement" (see Figure 8.3). This shows what mature or good really looks like from an observable point of view. The point of this is not to get external consultants to come in and tell the organization what the behaviors should be, but to use the framework to articulate these behaviors in a language that resonates with the organization. This also becomes a very convenient way for leaders to articulate expectations around the display of behaviors.

Let us examine the second CI behavior of "Developing and Implementing Solutions" across the multiple levels of an organization, as outlined in Figure 8.3. At the team member level, members and their peers identify opportunities to eliminate waste and improve processes from gaps in current performance using the eight-wastes model or team/customer feedback. They then generate a problem statement, establish and validate root causes, brainstorm potential solutions, prioritize those solutions, and then implement them. The team leader or manager will undertake the same activities, but they will also help to translate ideas into actions. In addition, the team leader could prioritize with the team the improvement initiatives to be worked on, create time for team members to work on those initiatives, and track the results to ensure sustainability. As you move up through the leadership structure, it is more about creating a safe environment and a CI enabling culture, and role modelling problem solving using the standard approaches—for example, PDCA, A3, Root Cause Analysis, etc.

8.4.2 Defining a Measurement System to Assess Behaviors

With the behaviors defined at multiple levels, the next challenge is to develop a measurement system that is both repeatable and reproducible. A repeatable measurement system is where the same assessor will score the same behavior on separate occasions and give the same result. A reproducible measurement system is where two different assessors observe the exact same behavior at the same time and give the same score. The most common, and recommended, scoring system is based on the Leichhardt system, with scores awarded between one and five or one and ten. As the scoring system is based on discrete data, you could either run an attribute agreement analysis to ensure absolute accuracy or go with the

Continuous Improvement

Continuously improving & innovating what we do to make things simple and easy for our customers & each other.

	Team Member	Team Leader / Manager	Department / Executive Manager	Vice President/ Head of/ General Manager	Group President / Executive General Manager	Group Executive/CEO
Fosters a culture of continuous improvement	Explores new ways of doing things.	Proactively and constructively challenges the status quo	Proactively and constructively challenges the status quo, and encourages others to do the same.	Invites others to challenge the status quo and envision possibilities to find better ways to achieve results.	Encourages others to research new approaches that will optimise results for the business/ division.	Encourages others to research new approaches that will optimise results for the organisation.
Develops & implements solutions	Identifies opportunities to eliminate waste, improve and or simplify processes and raises with Leader/Manager as appropriate.	Translates ideas into actions, considering both downstream and upstream impact.	Guides team to implement simpler ways of working considering downstream and upstream impact.	Provides guidance to the team on where to focus productivity initiatives.	Makes decisions about which improvement initiatives to resource.	Makes decisions about which improvement initiatives to resource.

Figure 8.3 Example CI Behaviors Across Different Levels of the Organization.

236 ■ *Why Bother?*

	Team Member	Team Leader / Manager	Department / Executive Manager	Vice President/ Head of/ General Manager	Group President / Executive General Manager	Group Executive/CEO
		Ensures improvement initiatives will deliver value before developing solutions.	Ensures improvement initiatives will deliver value before developing solutions.	Measures progress to ensure improvement efforts lead to positive and sustainable impact.	Measures progress to ensure improvement efforts lead to positive and sustainable impact.	
Lives Continuous Improvement	Understands Organisations Group's Productivity Standards and applies them constantly.	Understands Organisation Group's Productivity Standards and applies them constantly.	Understands Organisation Group's Productivity Standards and applies them constantly.	Champions Organisation Group's Productivity Standards; reinforces the need to constantly review and optimise the business.	Advocates the Productivity Standards to the business/ division and Productivity as an organisational capability.	Advocates the Productivity Standards to the organisation and Productivity as an organisational capability.
	Where applicable, follows current Standard Operating Procedures.	Invests time in monitoring adherence of team members to agreed Standard Operating Procedures.	Actively monitors performance and output across the team and reduces variation.	Constantly looks for ways the business can be more productive.	Constantly looks for ways the business/division can be more productive.	

Figure 8.3 Cont.

simple method of piloting with experienced assessors and recalibrating the measurement system definitions. We recommend the latter.

Let us explore what that could look like for a team member under Developing and Implementing Solutions. The overall behavior is described as "Identifies opportunities to eliminate waste, improve and or simplify processes and raises with Leader/Manager as appropriate."

On a Leichhardt scale, it could look something like that shown in Figure 8.4 across the different levels of maturity.

So how do you start to define this scale? Well, Figure 8.5 is a set of descriptors that should be used for each discrete score. The lowest score is where there is no evidence (none) or not applied, or ad hoc application, and/or not defined. Then, as you progress through the levels it could increase in numbers and lower variation. For example, Level 1 is no one is trained on problem-solving (less than five percent); Level 2 is a few people (five to twenty percent); Level 3 is some people (twenty to forty percent); Level 4 is most people (forty to seventy percent); and Level 5 is all people (over eighty percent).

Using consistent descriptors helps the assessors to measure behaviors consistently. It is important to understand what evidence is required to help the assessors to develop a score. For example, people leaders in interviews say that they empower their people to solve problems. What the assessors will have to then do is look for actual evidence within the organization of what is being talked about in the interviews. For example, in team meetings, are the team members putting forward improvement ideas or are they being prompted? Do they feel safe or engaged enough to suggest their ideas for improvement?

Once the measurement scale is defined for each desired behavior that is being assessed, you have a measurement system. The specific scoring of using only whole numbers, half marks, or even quarter marks gives the opportunity to recognize achieving some elements of the next level but not enough to give a full mark. It is important that assessors are well trained in using this high level of discretionary scoring measurement system.

8.5 Why Is It Important to Get Consistency Across Assessors?

With a measurement system now clearly defined and tested, the biggest challenge becomes the reproducibility between assessors. It is important to select assessors with the required traits and then train and develop them.

238 ◾ *Why Bother?*

Level 1	Level 2	Level 3	Level 4	Level 5
Team Members typically do not identify or raise improvement ideas	Few team members occasionally identify improvement ideas	More team members regularly identify improvement ideas.	Most Team members regularly identify improvement ideas most days	All Team members proactively and frequently identify improvement ideas
Problem solving is not performed.	Identification method to identify root causes, develop solutions, and follow up on the results.	A systematic methodology to solving problems is introduced, and some team members are have applied the methodology	The problem solving methodology is applied most of the time.	All levels apply the the CI methodology all of the time
Team Members lack the required problem-solving skills.	Most problems are solved in a reactive, 'firefighting mode'. A few team members in the organisation are trained in root cause problem solving.	Classroom training on problem solving exists, but is inconsistently applied	Most team members have been trained and coached in problem solving, and is regularly applied	All Team members are fully trained and actively involved in solving problems and implementing solutions.

Figure 8.4 Example Measurement Scale for Team Members on CI.

Level 1	Level 2	Level 3	Level 4	Level 5
None	A Few	Some	Most	All
Ahdoc	applied inconsistently	Sometime used	Regularly used	Consistently used
Not Defined	Defined but not understood	Understanding flowing through levels	good understanding at all levels	part of culture/ways of work

Figure 8.5 Descriptors to Help Score Definitions.

8.5.1 Selecting the Best Assessors

The key characteristics for selecting assessors are:

1. **Practical experience**—senior operational leadership experience (desirable), previous behavioral assessment experience, Lean/Six Sigma/transformation experience.
2. **Trusted leader**—can stand up to other leaders, have tough conversations, and provide recommendations, and has mandatory senior people leadership experience.
3. **Curious**—growth mindset; the maturity assessment is a learning experience.
4. **Systems thinker**—can both be in the process and look beyond individual items, can synthesize common themes, and make connections.
5. **Empathetic**—can walk in the shoes of those going through the assessment, not to judge but rather to inspire.
6. **Clarity**—can articulate and pitch recommendations to various levels of leadership.
7. **Disciplined**—can leverage and improve the standard work to ensure the best results.

There are a few extra characteristics for the section of the lead assessor:

8. **Senior leadership experience**—mandatory to have had senior leadership experience across multiple business units and, ideally, organizations.
9. **Humble**—willing to learn from other assessors.
10. **Coach**—strong, demonstrated ability to coach and develop assessors in the team.

The most challenging question is whether to recruit externally, use management consultants, or develop internally. Each has its strengths and weaknesses.

Recruiting external assessors with significant Lean and transformation experience is a great start, as they bring both experience and fresh eyes to the assessor development pool. However, they also lack specific organizational knowledge and credibility.

Bringing in management consultants can be great in the initial setup and program launch provided care is taken to ensure knowledge transfer to internal staff. External consultants tend to want to develop a future revenue stream by keeping intellectual property control so do not transfer assessment capability to the organization. If engaging external firms, confirm that there is a strong clause on capability transfer to ween the organization off dependency on external consultants. However, it is advantageous to bring in someone external to help design the assessments program and initial development of internal assessors.

Developing assessors internally involves having a full-time core team and embedded assessors from within the business. Recruiting internally has the strengths of strong relationships and organizational knowledge and context but risks a simple lack of knowledge of or experience in assessing "behaviors" and not use of tools—this is also a challenge with external candidates and management consultants. However, having a large, central team can be seen as an excessive overhead, and often, fully embedded assessors do not conduct enough assessments to maintain their competency or mature the program. The optimal approach of all three is a small core team that continually coaches the capability of not only the embedded business assessors but also a wider group of regional/embedded assessors.

It is important to develop the recruitment and onboarding of assessors against the above characteristics. As with all recruitment, it is important to have a structured development program for the assessors.

A mistake that organizations make—to provide a development opportunity—is to purposely select internal people without significant leadership experience or, in some cases, with no people leadership experience at all. This will have a significant impact on the credibility of the overall assessment program. That is why senior leadership experience is mandatory and should never be negotiable, especially for the lead assessor.

Another great opportunity to help develop the maturity of the transformation program being assessed is to include senior leaders from the organization to participate in individual assessments. There are several reasons this is successful:

1. It shows organizational commitment to have senior leaders create time to attend assessments.
2. It gives strong credibility to the report back of the assessment team, as they can give very specific context and linkage to the broader business.
3. It gives the senior leader some strong practical experience in understanding behavioral assessments and establishing KBIs.

The role modelling of senior leaders cannot be underestimated.

8.5.2 Calibrating Understanding and Scoring Across Best Practice

As the assessment progresses it is important for the assessment team to continue to review the information and evidence being captured and, more importantly, gaps in evidence.

It is important to have some sort of tool to allow the lead assessor to review the variation across the assessors and use this as a data point for assessor coaching and development.

The process of calculating the calibrated score is to have a discussion between the assessors and the assessment lead. The best approach is for each of the assessors to give a simple rationale of their score based on what they saw or did not see. The scoring should not be a consensus; the assessment lead should try to get an agreement between the assessors but should have the final decision on the score.

An important part in assessing the behaviors is not just the use of the words of the definition but to look also at the frequency, intensity, duration, and depth when using the behavioral assessment scale.

Frequency—How regular the behavior is being performed, that is, hourly, daily, weekly, monthly, or annually?
Intensity—The energy of the display of the behavior. Are people really passionate about the behavior or just doing it out of compliance?
Duration—How long this behavior has been in place. Is it recent, the last couple of weeks/months, or has it been around and improving for a reasonable amount of time?
Depth—How well is this behavior being seen through all levels in the organization and across the departments?

New Trainee Assessor	Developing Assessor	Assessment Lead	Lead Assessor
• 1st-time participant as an assessor
• Observes the process paired with experts
• Sits in on interviews, huddles, readout
• Learns how to conduct effective interviews
• Learns how to translate observations into scores using the assessment maturity grid and tool
• Provides quotes and calibrates with team
• Participates in readouts, adds color as needed	• Participated in at least 1 previous maturity assessment
• Demonstrates skills and behaviors that position assessor for certification (unbiased, appreciative, inclusive, etc.)
• Sets up the agenda of the Assessment
• Leads interviews with expert assessor
• Scores across all 4 disciplines
• Leads calibration discussion with an expert assessor
• Learns to produce the report
• Leads elements of the readout
• Scores and approach are mostly aligned with experts	• Calibrated with assessment methodology and scoring owners, and potentially external
• Leads the assessment team through the assessment process: agenda set up, walkthrough, interviews, scoring, calibration, synthesis, and readout
• Trains assessors
• Engages with external benchmarks	• Coach and Develop Assessment Leads
• Reviews improve assessment process and standard work.
• Coach Senior Manager guest assessors

Figure 8.6 Potential Assessor Learning Journey.

A simple way is to have a spreadsheet or tool that collates by ideal behavior the individual scores given by each assessor. Once the agreed team score is compiled, then the variation across the assessors to the calibrated final score can provide some great insight overtime on the maturing of the assessor's capability. Quite often, a standard deviation or number of scores greater than 0.5 from the calibrated final score can be used as a measure of variation.

8.6 Why Develop the Assessors?

Now is the time to develop the assessors towards continually improving the assessment program maturity and their own skill levels

Figure 8.6 shows a potential learning journey for the development of internal assessors from trainee through to qualified assessor and further on to assessment lead and lead assessor.

8.7 Key Takeaways

1. You need to care deeply about the robustness and consistency of assessments if they are to have credibility across the organization.
2. Drive consistency across the assessment team, especially when onboarding new assessors, through a standard training and calibration methodology.
3. It is important to have robust conversations during score calibration to provide the most value in feedback to the customer.

Chapter 9

Why Bother Having a High-Level Roadmap to Deploy?

Chapter Summary

A high-level visual roadmap is an invaluable planning and communication tool. It is a quick way to show the organization what is involved. It enables the creation of detailed action plans and milestone planning and provides a framework for initial design and ongoing improvement of the assessment system. Each organization has a unique context with different starting points, different cultures, different business languages, geographic complexities, and organization structure. As such the roadmap in this chapter is meant as illustrative and we encourage people to adapt it to their own specific context and requirements.

In the preceding chapters, we have described the detailed roadmap activities required to develop your CI assessment system. A key point to remember is that the development of the assessment system is never finished. It must continuously evolve and improve with ever-increasing understanding and expectation of what it is possible to achieve.

A high-level visual roadmap is an invaluable planning and communication tool. It is a quick way to show the organization what is involved. It enables the creation of detailed action plans and milestone planning. It provides a framework for initial design and ongoing improvement of the assessment system. Each organization has a unique context with a different starting point, culture, business language, geographic complexities, and organizational structure. As such, the roadmap in this chapter is illustrative and we encourage people to adapt it to their own specific context and requirements.

The roadmap below shows two high-level PDCA cycles. These, however, are not the whole picture, as many of the PDCA phases will involve standalone "mini" PDCA cycles. One example is the communications pack. We need to plan this (Plan), create it (Do), get feedback from test groups in several areas across the organization (Check), and implement feedback and redesign the planned approach as required (Act). We encourage you to develop your own "mini" PDCA cycle within each of the main elements shown in Figure 9.1.

Some of the important questions to ask at each phase of the first PDCA cycle are given below:

Plan

Do we have consensus and senior leadership understanding on the purpose?

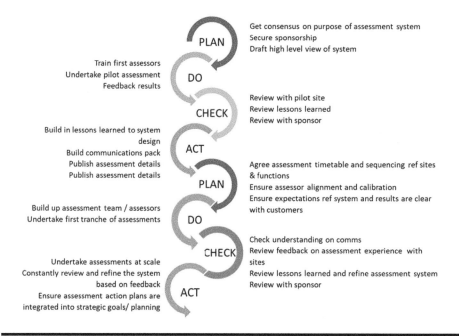

Figure 9.1   The Assessment System Development Roadmap.

Do we have proactive sponsorship at the right level?
Is the system design clear and at the right level of detail for senior leaders?

Do

Do we have a structured and visible process and criteria for the selection of assessors?
Do we have customer pull from the pilot area?
Are expectations clear with the internal customers on the outcome and actions required from the feedback?

Check

Have we spent sufficient time obtaining and understanding customer feedback and ideas for improvement—for example, formal Voice of the Customer activity?
Do we have a clear process of capturing and evaluating improvements?
Have we engaged the sponsor sufficiently so that they are a vocal advocate?

Act

Do we have a visible process for tracking the implementation of improvements?
Do we have a visible process for getting feedback on the communications pack?
Do we have a clear plan for checking if the communications pack has been correctly understood?

Some of the important questions to ask at each phase of the second PDCA cycle are given below:

Plan

Do we have a clear and visible process for sequencing of sites and functions?

Do we have a clear and visible process for assessor alignment—for example, a skills matrix?

Have we confirmed expectations with key stakeholders?

Do

Is the assessor role positioned as a highly valued position that will lead to future opportunities?

Do we have clear "mini" PDCA cycles in place with assessor teams and customers for each of the assessments?

Are assessment schedules sequenced to support input to strategic planning cycles?

Check

Were the site expectations of the assessed sites/functions aligned to the communications pack?

What can we learn from the Voice of the Customer and senior sponsor from their feedback on the assessment system and process?

Do we have a clear and visible process for responding to feedback, and evaluating and incorporating lessons learned?

Act

Do we have a clear and visible process for evaluating action plans and their implementation?

Do we have a clear communication plan to explain the incorporation of lessons learned?

Do we have an effective, ongoing Voice of the Customer process?

At a high level, some key features of a good assessment system that need to be continuously reviewed are:

Is it still highly valued by the leadership teams across the organization?
Is it still driving action plans for further improvement?
Is it still supporting the never-ending pursuit of ideal behaviors?

If the answer to any of these questions is not a resounding yes, then the system needs to be reviewed and amended accordingly. A key thing to remember is that the answer to these questions needs to come from the customers, not the CI team.

9.1 Key Takeaways

1. A high-level visual roadmap is an invaluable planning and communication tool.
2. A multi-cycle Plan, Do, Check, Act approach is the most effective way to design, deploy and continuously develop the assessment system.
3. Create your own visual roadmap reflecting your context and business language.

Chapter 10

Why Bother Doing Assessments?

Chapter Summary

The CI maturity assessment needs to be integrated into the strategic business planning cycle so that outputs and opportunities can be incorporated into forward planning. They should not be a stand-alone activity to business planning but rather a key check on progress and a key input for consideration in action planning. This chapter gives a summary of the key bullet points for the why, the what, and the how of developing and undertaking CI maturity assessments.

Each organization has a unique context with different starting points, culture, business language, geographic complexities, and organizational structure. As such, a lot of thought has been given to the approach, design, and content of the assessment; customizing to local requirements is critical to success.

Our intention in the preceding chapters of this book has been to share the why, what, and how of CI maturity assessments. We believe they are an essential element to developing and sustaining a CI culture. They do not work as standalone tools or systems, but instead need to be integrated into the overall transformation program and used as a key component of the strategic "check" on progress and lessons learned. They provide a high-level strategic Key Behavioral Indicator (KBI) and are a powerful form of recognition. Below are some key bullet points summarizing the why, what, and how.

10.1 The Why

- Essential to understand what is working well and why so that best practices can be built on and shared.
- Essential to understand the opportunities for further improvement and what might not be working as well as expected in the approach to CI.
- It provides a strategic level KBI and is a powerful form of recognition.

10.2 The What

- An assessment system customized to your own context, business language, and culture will provide far more value than a generic, off-the-shelf assessment.
- It must be based on assessing behaviors and not just tools or systems, but rather an integration of all these.
- It must be owned and valued by the business leadership teams.

10.3 The How

- All feedback must be constructive, and the system designed to demonstrate that it is intended to help people be even more successful.
- The assessment system needs to constantly evolve in a continuous Plan, Do, Check, Act cycle.
- Undertake assessments with humility. There is always something to learn.

Like everything else in CI, what is most important is not what you do but how you do it. If the assessment system is perceived as an audit that is designed to check up on people or a tool for centralized micro-management, it will fail to support cultural transformation. Indeed, it is more likely to drive poor behaviors and create more problems than benefits.

We recommend to always approach designing and running a maturity assessment system with a large dose of genuine humility. Listen to people's concerns and openly discuss and address these. It is not necessary to design the perfect system from day one. Getting something up and running that people are conformable with is more important than letting initial perfection of the system potentially stop improvement.

We have tried to show that the system needs to continuously evolve and that a PDCA approach to every aspect will create the best chance of success. Developing a CI culture is a never-ending journey, where the destination will keep changing as your understanding deepens. The maturity assessment system needs to adapt to this and support the journey at all stages.

Several Shingo Prize-winning organizations have responded to the award with the statement that it was fantastic to receive the reward and they really valued the recognition it gave to their people. However, the most valuable thing for them was the insights from the examiners' feedback. No matter how highly they had been rated, they knew there was always more to go for. This humility and relentless pursuit of improvement are essential to creating a sustainable culture of continuous improvement.

10.4 Key Takeaways

1. The CI maturity assessment system needs to be integrated into the business planning cycles to function as a check on progress and inform future business planning.
2. The assessment results can be used as a high-level strategic KBI for the whole organization.
3. The assessments provide a strategic-level check on the progress and sustainability of the CI culture.

Appendix

Version 1	
Customer Focus	
1	Our customers (both internal and external) are clearly identified throughout our business.
2	There are documented processes throughout our business to regularly seek feedback on what our customers value, and how well we are delivering this.
3	Each team in our business has developed a formal Customer Value Proposition, defining how they deliver value to the customer.
4	In our business, we regularly measure and track metrics that impact customer satisfaction eg. number of errors, complaints, response times etc.
5	In our business, each team regularly reviews the metrics that impact customer satisfaction to identify opportunities for improvement.
6	We work with our internal business partners to help them understand what our team does and how we add value to our customers
Process Management	
7	In our business we have developed process maps of our core processes that truly reflect the work completed within each process.
8	In our business, standard operating procedures are readily available to users.
9	In our business, we regularly measure and report on the performance of our key processes using Visual Management Boards.
10	Our Visual Management Boards adhere to the Group standard of 3 core components: Performance Measures, Continuous Improvement and Actions.
11	We work closely with upstream and downstream teams that are part of the end-to-end process.
12	In our business, the handover points of our processes to and from other departments are well defined and clear to everyone.
13	Within our business, each core process has a process owner who regularly reviews process performance, identifies opportunities for improvement, approves changes to the process etc.

14	In our business, each core process has a process governance body, led by the process owner, which brings together all the stakeholders that impact the performance of the process.
15	Accountability is clear at all levels of our business.
Productivity Capability	
16	In our business, there is a well-defined Productivity program for everyone to identify and get involved in opportunities for improvement.
17	In our business, everyone has received training on a range of different Productivity and problem solving tools and techniques eg. Productivity Essentials, and the Four Productivity Habits.
18	Other sources of information on Productivity tools and techniques (incl. how to use them) are readily accessed by everyone when required, to complement the formal training programs (e.g. Intranet tool guides, coaching, Chatter etc).
19	Progress on local Productivity initiatives, incl. the number implemented and the benefits delivered, is highly visible and regularly communicated at all levels of our business.
20	Productivity initiatives are making a definite improvement to the way the business works and enabling us to meet our overall Productivity targets.
21	A process exists at all levels for teams to establish projects and for actions to be resolved locally.
22	In our business we are capable of effectively applying a range of problem solving and process improvement techniques.
Productivity Behaviours & Mindset	
23	The Four Productivity Habits (Standard Operating Procedures; Visual Management Boards; Huddles; Continuous Improvement) are embedded throughout our business.
24	In our business, Huddles are prioritised and fully attended by team members.
25	Continuous Improvement is a daily habit for everyone and systems are in place to enable this behaviour (ie PDCA).
26	In our business, when people work together with people from other teams on Productivity opportunities, they work cooperatively and try to find the right solution for the whole organisation.
27	We ensure everyone in our business has a Personal Development Plan, identifying training courses, coaching, leadership training etc to enable us to build and improve our skills and capabilities, including Productivity skills.
28	We ensure that individuals in our business who make a significant contribution to improving performance are publicly acknowledged and recognised.
29	In our business, we ensure performance appraisals include Productivity measures, and individuals are evaluated and rewarded on them.

	Productivity Leadership
30	Leaders within our business consistently champion and promote the importance of productivity improvement to their teams.
31	Leaders within our business are role models for constantly improving through visibly applying productivity improvement tools to their own management activities.
32	Leaders within our business are actively engaged in performance improvement programs and projects, ensuring the right level of resources, skills and funding are available.
33	Leaders in our business adopt a coaching style, helping their direct reports and teams think through and solve problems for themselves rather than telling them the answer.
34	Leaders in our business encourage team members to be resilient and persistent in Productivity initiatives, even if initial attempts are not successful.
35	Across our business, an appropriate balance of time is spent on strategic improvement and day job activities.

	Version 2
	Customer Focus
1	Our CVP is established and used on a regular basis to inform our decisions.
2	There are documented processes that are regularly followed to gather feedback from our internal and external customers.
3	In our business, each team regularly reviews the metrics that impact customer satisfaction to identify opportunities for improvement.
	Process Management
4	In our business we have developed process maps of our core processes that truly reflect the work completed within each process.
5	In my business we regularly check that the documented SOP reflects what people are doing.
6	All SOPs are reviewed at least once per year, in addition to changes as a result of CI.
7	There are clearly defined owners for each of the processes that I use.
8	Everyone in my business regularly attends at least one huddle per week.
9	Our huddles evolve and continue to add value to each person in the huddle.
10	We work closely with up and down stream teams to understand and improve the process.
	Productivity Capability
11	In our business there is a well-defined Productivity program and everyone is encouraged to identify and get involved in opportunities for improvement.

12	In our business we use a standard Productivity framework (eg PDCA, DMAIC) to solve problems.
13	Progress on Productivity initiatives / continuous improvement including the number implemented and the benefits delivered, is highly visible and regularly communicated at all levels of our business.
14	In our business we are encouraged to challenge existing ways of working.
Productivity Behaviours & Mindset	
15	I believe Productivity is about making things simple and easy for customers and each other.
16	I know how my work contributes to achieving the CVP.
17	I understand how the metrics on our VMB measure our progress to achieving the CVP.
18	I actively use process maps eg to understand where I contribute to the process, or to help with CI.
19	I know which productivity tools I am expected to use to solve problems.
20	I use productivity tools to solve problems.
21	I measure and record the benefits that my CI's produce.
Productivity Leadership	
22	Managers and leaders in my business actively champion Productivity by completing Gemba walks.
23	Managers and leaders in my business role model Productivity by using the Productivity tools to solve problems.
24	Managers and leaders in our business use an enquiry-based coaching style where they help people understand and solve problems for themselves rather than telling them the answers.
25	Leaders in our business show that they value Productivity capability through encouraging development and utilising their certified Green / Black belts.
26	*Leaders in our business provide clear, consistent and frequent communications about direction and expectations. Leaders in our business provide clear, consistent and frequent communications about direction and expectations.*
27	An action plan was implemented after the last survey.

Index

Note: Page numbers in **bold** indicate tables; those in *italics* indicate figures.

4+1: Embedding a Culture of Continuous Improvement in Financial Services (Jones, Butterworth, and Harder) 1, 36, 41, 49, 186
5S 34; audit extract example *34*

accountability: assessment report 227; Bakkavor Desserts 205; Commonwealth Bank of Australia 188, 193; financial services organization case study 65; NBN Co. 162; role modelling ideal behaviors 53, 57; undesired behaviors, dealing with 53
action planning workshop 226–227; assessment team roles 229, 233; designing a behavioral management system 101
Agile 103
Airbus Australia Pacific (Airbus (AP)) 177–185
Alder, Jon 104
alignment of assessments, assessors, and calibration 221–243; assessment team roles 227–233; assessor development 243; importance of consistency 237–243; maturity 222; process 222–227; reasons for consistency 233–237
approval of assessments 222–224
Argyris, Chris, *Theory in Practice* (with Schön) 83, *84*
assessment coordinator 232–233

assessment team roles 227–233
assessors: Airbus Australia Pacific 182–183; calibration 241; Commonwealth Bank of Australia 191; consistency across 237–243; development 229, 230, 232, *242*, 243; measurement system 237; NBN Co. 163, *164*; number of 95, 96–97, 100; roles 227, 229–233; selecting the best 239–241; self-assessment 227; virtual assessments case study 212–213, 215, 218
attention 40–41
audits: Airbus Australia Pacific 185; assessments versus 3–6, **4**; designing a behavioral assessment system 72; Panalpina logistics division 176

Bakkavor Desserts 198–212
behavioral assessment 33–68; designing a strategic level system 71–101; purpose of 35–58
Behavioral Deployment System 22, *23*; development 109–134, *110*; STMicroelectronics 112–134, *114–123*, *125*, *127–131*, **132–133**
Behavioral Playbook 65, *66*
behaviors 15–17, 30–31; assessment *see* behavioral assessment; building the expected 60–63, *61*; Commonwealth Bank of Australia 191–192, 195–196; and conversation 140, *142*; and culture

259

106–107; defined 106–107; defining and linking with systems 18–25; defining for assessment 233–234, *235–236*; Enterprise Alignment principles **28**, **29**, **30**; financial services organization case study 59–67; Gemba walks **45**; ideal *see* ideal behaviors; KBIs *see* Key Behavioral Indicators; measurement 107–109, 234–237, *238–239*; NBN Co. 161–162; purpose and systems as drivers of 16, 18–25, *21*; undesired 52–53, 74–75

Bessant, John, *High-Involvement Through Continuous Improvement* (with Caffyn) 83, *84*

best practice 73–74, 227; Airbus Australia Pacific 184; assessment team roles 229, 233; calibration 241–243; designing a behavioral management system 101; NBN Co. 166

blame, removing 53

brain, and habit formation 37–42

Brain that Changes Itself, The (Doidge) 38

Brun, Philippe 124

Business Process Excellence (BPE, NBN Co.) 164, 166

Butterworth, Chris: *4+1* (with Jones and Harder) 1, 36, 41, 49, 186; *The Essence of Excellence* (with Hines) *see Essence of Excellence, The*

Caffyn, Sarah, *High-Involvement Through Continuous Improvement* (with Bessant) 83, *84*

calibration 225, 241–243; assessment team roles 228, 230, 231

Cardiff Business School 168–169

Carrots and Sticks Don't Work (Marciano) 58

Champions 9–10

clarity: of assessor 239; characteristics of good Sponsors 11; on purpose 73–75

Coaches/coaching: by lead assessor 239; and Sponsor success 13

collaboration 73–74, 91

commitment, as good Sponsor characteristic 11–12

Commonwealth Bank of Australia (CBA) 185–198; 3Cs Productivity Program 186, *186*, 188; accreditation system 187–188, *187*, **188**; Commway/ Productivity program 186

communication: Bakkavor Desserts 205–208; Commonwealth Bank of Australia 197–198; designing a behavioral assessment system 97–98; Gemba walks **45**; NBN Co. 163; virtual assessments case study 215; *see also* conversations; talk

conformance phase 149–152, *150*, **151**

consistency: as good Sponsor characteristic 12; importance of 237–243; of processes 37; reasons for 233–237; *see also* alignment of assessments, assessors, and calibration

Constancy of Purpose principle 28–29, **29**

conversations: continuum of conversation model 144, *145*, 153; reasons for focusing on 137–157; *see also* communication; talk

core operating systems *106*

COVID-19 pandemic 212–213, 216, 217

credibility of assessment 221, 233; internal assessors 240; senior leaders' participation 241

criticality, as good Sponsor characteristic 11

Cultural Enabler principles 57

culture 1–2; and behaviors 106–107; Commonwealth Bank of Australia 187; conversation 138–140; fire-fighting 143; Gemba walks 49; managing 35; respect 57; systems reviews and improvements 50

curiosity, assessor 239

Customer Scorecard (Bakkavor Desserts) 201

customer value 36–37; designing a behavioral assessment system 76; Enterprise Alignment 27–28, **28**

customers: of assessment system 76; focus on (Commonwealth Bank of Australia) 192–193; voice *see* Voice of the Customer

Deming, W. Edwards 22
descriptors for scoring 237, *239*
designing a behavioral assessment system 71–101; Airbus Australia Pacific 181–182; Bakkavor Desserts 202–208; Commonwealth Bank of Australia 192–197; NBN Co. 162–163
discipline, assessor 239
disruptive behavior 86
documentation for assessment system 96–97
Doidge, Norman, *The Brain that Changes Itself* 38
duration of the assessment 95, 100

efficiency of processes 37
embedded assessors *see* permanent and embedded assessors
empathy, assessor 239
employee retention 43
empowerment: developing a Behavioral Deployment System 111; measurement system 237; STMicroelectronics 114, 115–116, *117*
engagement: Bakkavor Desserts 205; financial services organization case study 65; maturity level pre-requisites 89; and recognition 42, 43; respect 58; role modelling ideal behaviors 55
Enterprise Alignment principles 26; Constancy of Purpose 28–29, **29**; Customer Value Creation 27–28, **28**; Think Systemically 29–30, **30**
Enterprise Excellence 25–26, *26*, 104; Bakkavor Desserts 199; Commonwealth Bank of Australia 191; conversations 138–140; designing a behavioral assessment system 76, 77; and the disconnected bridge 104, *105*, 113; Panalpina logistics division *174*
Essence of Excellence, The (Hines and Butterworth): behavioral deployment and behavioral-based recruitment 87, 112; culture of continuous improvement 20–22; Gemba walks 46, 47; ideal behaviors, defining 23; recognition 42
experience, assessors 239
external assessors, strengths and weaknesses of 240

feedback 227; assessment team roles 229, 230, 232; Bakkavor Desserts 203, 205, *206*, 208; reports 96–97; virtual assessments case study 219; workshops 100, 101
financial services organization case study 59–67
financial targets: Bakkavor Desserts 198–199; Commonwealth Bank of Australia 188–189; maturity level pre-requisites 89
fire-fighting culture 143
follow-up of assessment 226–227
framework for behavioral assessment system 76–77
frequency of assessments 93; NBN Co. 163

Gemba walks 44–49; assessment process 222, 225; assessment team roles 228, 230, 231, 233; Bakkavor Desserts 203; designing a behavioral assessment system 95; financial services organization case study 60, 65; NBN Co. 166; six-step approach *46*; themes **45**; virtual assessments case study 212, 216–217
governance, agreeing timing with 224
group size for assessment 78, 92–94
guest assessors 231–232

habit formation 37–42
Harder, Brenton, *4+1* (with Jones and Butterworth) 1, 36, 41, 49, 186
Hearing the Voice of the Shingo Principles (Miller) 57
High-Involvement Through Continuous Improvement (Bessant and Caffyn) 83, *84*
high-level visual roadmap 245–249, *246*

Hines, Peter: *The Essence of Excellence* (with Butterworth) *see Essence of Excellence, The*; Panalpina logistics division 171

how, the 252–253

huddles: Commonwealth Bank of Australia 188, 197; designing a behavioral assessment system 100; habit formation 41, 42; NBN Co. 166; peer role modelling of ideal behaviors 56–57

humility: Commonwealth Bank of Australia 191; credibility of assessment 221; developing a Behavioral Deployment System 110; lead assessor 239; role modelling ideal behaviors 53, 56

ideal behaviors 30–31; designing an assessment system 72, 73–74; examples 23, 73–74; and ideal results 16, *16*, 17–18; principles, role of 16–17, 25–26; purpose and systems as drivers of 16, 18–25, *21*; reinforcing 35–58; role modelling 53–57, 166, 184

impartiality, and credibility of assessment 221

incremental step change phase 152–154, *152*, *153*, **154**

innovation 27, 42

internal assessors 229–230; strengths and weaknesses 240

interviewees: designing a behavioral assessment system 94–95, 100, 101; virtual assessments case study 219

interviews: assessment process 222, 225; assessment team roles 228, 230, 231, 233; designing a behavioral assessment system 94–95, 99–100; virtual assessments case study 215, 217, 218, 219

Jones, Morgan, *4+1* (with Butterworth and Harder) 1, 36, 41, 49, 186

journey to sustainable continuous improvement culture 139–140, *139*

Kerr, James 61, 65

Key Behavioral Indicators (KBIs) 17, **108**, 251–252; financial services organization case study 61, 63, 65; Gemba walks 49; and Key Performance Indicators, relationship between 17, *17*, 107–109, *108*; level 3 framework *134*; reasons for defining 103–135; role modelling ideal behaviors 57; senior leaders' participation in assessment 241; sponsorship 9, 13; STMicroelectronics 118–134, *120–123*, *127*, *130–131*, **132–133**; systems and behaviors 20, 24

Key Performance Indicators (KPIs) 17, 103; Bakkavor Desserts 200; designing a behavioral management system 100; financial services organization case study 61, 65; Gemba walks 46; and Key Behavioral Indicators, relationship between 17, *17*, 107–109, *108*; lag **108**, 126; lead **108**; NBN Co. 160; Panalpina logistics division 176; sponsorship 9, 13; STMicroelectronics 113, 126, *129*, 134; systemic thinking 29; systems and behaviors 20, 24

Kirkpatrick model 134, *135*

Lahy, Andrew 171, 175

languages for assessment 78

lead assessor 227, 228–229; calibration 241; selecting the best 239

Leader Playbook 65, *67*

leading indicators 17; team size for assessment 94; *see also* Key Behavioral Indicators

league tables 91

Lean 103; Airbus Australia Pacific 177–181, *182*, *183*, 184; Bakkavor Desserts 201; Champions 9; designing a behavioral assessment system 76; maturity measurement 222; Panalpina logistics division 169, 171–172, *172*, *174*; Sponsors 10; STMicroelectronics 112–113, 116–118, 123–124, 126; virtual assessments case study 212

Lean and Play (STMicroelectronics) 116–118, *120*
Learning and Development system *23*; purpose statement 22, *50*; review and best practice sharing 101
Learning to Lead at Toyota (Spear) 139, 152
Legacy (Kerr) 61, 65
Leichhardt system 234–237
Lencioni, Patrick 124
lessons learned reviews 98, 100
local-level assessment 78, *79–83*
LogEx (Panalpina logistics division) *170*, 170–177, *174*
Look, Listen, and Learn walks *see* Gemba walks

magic wand question 100, 226; Commonwealth Bank of Australia 197
management consultants, strengths and weaknesses of 240
managerial relationships, and recognition 43
Mann, David 47
Marciano, Paul 58
maturity example *223*
maturity insight meetings (MIMs, Commonwealth Bank of Australia) 190
maturity levels: Bakkavor Desserts 200–201, *202*; Commonwealth Bank of Australia 187–188, **188**, 189–191, 197–198; designing a behavioral assessment system 83–89, *84*, *85*; NBN Co. 160–161, *160–161*; observed behaviors at different **141**
maturity roadmap 222
maximum team size for assessment 78, 93
McGill, Michael E., *The Smarter Organization* (with Slocum) 83, *84*
Medina, John 37
meetings: assessment process 222, 225; assessment team roles 228, 230, 231, 233; designing a behavioral assessment system 96, 100; maturity insight (Commonwealth Bank of Australia) 190; stand-up *see* huddles; virtual assessments case study 216
Miller, Robert Derald 57
"Mind the GAP" 40–41
minimum team size for assessment 78, 93
model for behavioral assessment system 76–77
Morrow, Bill 160

NBN Co. 159–167
negative behavior 86
Net Promoter Score 65
neuroplasticity, and habit formation 38–41
nudges 36
Nummi 57–58

O. C. Tanner Institute 42
observations: assessment process 225; Bakkavor Desserts 205, *208*; designing a behavioral assessment system 95, 99; virtual assessments case study 216, 217, 218–219
Operational Excellence 103; Bakkavor Desserts 201; Panalpina logistics division 169–177, *170*

Panalpina logistics division 167–177
passionate behavior 87
passive behavior 86
PDCA *see* Plan, Do, Check, Act
peer role modelling of ideal behaviors 56–57
people development 111; *see also* assessors: development
permanent and embedded assessors 229–230; strengths and weaknesses 240
Phillips, Jack and Patti 134
Pia Caraccia, Maria 172–175
Plan, Do, Check, Act (PDCA) 5–6, *5*, 105, 252, 253; designing a behavioral assessment system 72, 85, *98*, 98; high-level visual roadmap 246–249, *246*
planning the assessment 98–100

positive feedback, and habit formation 40–41
preparation for assessments 98–99, 222–224; assessment team roles 229–230, 231, 232–233; virtual assessments case study 214–215
principles-based approaches *104*, 104; and ideal behavior 16–17, 25–26
proactive behavior 87
process control phase 149–152, *150*, **151**
process for assessments: example *224*; reasons for 222–227
process management (Commonwealth Bank of Australia) 193–194, 195
process maturity 34; NBN Co. 160, *160–161*, 163, 164, 166, 167
productivity: capability (Commonwealth Bank of Australia) 194–195; leadership (Commonwealth Bank of Australia) 196–197; and recognition 42
purpose 4; clarity on 73–75; constancy of 28–29, **29**; designing a behavioral assessment system 72; as driver of behavior 16, 18–25, 50; statements 22, 24, 50, *50*, 75, 76

reactive behavior 87
readback session 226; assessment team roles 229, 230, 232
RealWear head-mounted tablets (HMTs) 213–218
recognition 42–43, 90
recruitment 58–59, *59*; of assessors 239–241; developing a Behavioral Deployment System 112; financial services organization case study 60–63, *62*, *64*
repeatable measurement systems 234
reports 227; assessment team roles 229, 230, 232, 233; compilation 226, 229, 230, 232
reproducible measurement systems 234
respect 57–58; Bakkavor Desserts 203; designing a behavioral assessment system 76; developing a Behavioral Deployment System 110; role modelling ideal behaviors 54, 56
results of assessments: Airbus Australia Pacific 183–184; Bakkavor Desserts *207*; NBN Co. 163, 164, *165*; role modelling ideal behaviors 55–56; *see also* feedback; reports
reviews 101
rewards 91; NBN Co. 163
Rock, David 37, 41
role modelling: of ideal behaviors 53–57, 166, 184; by senior leaders 241
Rother, Mike 139

SA Partners Lean Business Model 181, *182*, 184
safety issues: developing a Behavioral Deployment System 110; maturity level pre-requisites 90
SCARF model 41–42
Schein, Edgar 35
Schön, Donald A., *Theory in Practice* (with Argyris) 83, *84*
scope of behavioral assessment system 77–83
scoring: assessment team roles 228, 230, 231; defining a measurement system 234–237, *238–239*; process 225
Shingo, Shigeo 110
Shingo Institute: Airbus Australia Pacific 183; behavioral assessment scale **88**; *Best Ways for Manufacturers to Boost Employee Engagement* 43; Bronze medallion 83; Commonwealth Bank of Australia 188, 189, 191; customer value 28; duration of the assessment 95, 100; maturity levels 83–84, 86; Model *see* Enterprise Excellence; number of assessors 95, 100; Panalpina logistics division 168, 173–175, *174*; purpose and systems as driver of behavior 50; recognition 43; Shingo Prize (Gold medallion) 84, 90, 188, 253; Silver medallion 84; systemic

thinking 29; systems 18; Three Insights of Enterprise Excellence 16–17

Shingo Model *see* Enterprise Excellence

Shingo Principles 26–27, *27*; Cultural Enabler 57; developing a Behavioral Deployment System 109, 110; Enterprise Alignment 27–30; financial services organization case study 63, 65; respect 57

site-based assessment 77, 78

sitting in on meetings *see* meetings

Six Sigma 103; Commonwealth Bank of Australia 186

Slocum, John W., Jr, *The Smarter Organization* (with McGill) 83, *84*

Smarter Organization: How to Build a Business That Learns and Adapts to Marketplace Need (McGill and Slocum) 83, *84*

SoundWave 138, *142*, 142, 156–157, *157*; conformance phase *150*; high variation phase 143–144, *144*; incremental step change phase *153*, 154; reducing variation phase *147*

Spear, Stephen J. 139, 152

Sponsors 9–13; Airbus Australia Pacific 179; characteristics of good 10–12; Commonwealth Bank of Australia 188–191; NBN Co. 161, 167; success 12–13

standards, ignored 54–55

Step Back management practice (STMicroelectronics) 114–115, *116*

STMicroelectronics: Behavioral Deployment System 112–129, *114–123*, *125*, *127–131*, **132–133**; MAJU program 113–122, *114–123*

Strategy Deployment System 22, *23*, 109

stress, effect on the brain 38

success: elements of 12–13; measurement of 13

suppliers 82–83

Sustainability management practice (STMicroelectronics) 114, 115–116, *118*

SWOT analysis, Bakkavor Desserts 205

systems: constant review and improvement 50–52; defined 18; as driver of behavior 16, 18–25, *21*, 50; Thinking Systemically principle 29–30, **30**

systems-based approaches *104*, 104

systems thinking, assessor 239

talk: conformance phase 152; and continuous improvement 139; importance of 138; phase by phase 140–154; reducing variation phase 149; *see also* communication; conversations

target setting 92

team size for assessment 78, 93–95

Theory in Practice: Increasing Professional Effectiveness (Argyris and Schön) 83, *84*

Thinking Systemically principle 29–30, **30**

timing of assessment 224

tools-based approaches *104*, 104

Total Quality Management 169

Toyota 104, 110

Toyota Kata (Rother) 139

trainee assessors 231–232

translators 78

transparency: assessment report 227; credibility of assessment 221

trust: assessors 239; and recognition 43; undesired behaviors, dealing with 52

Value Stream Mapping (VSM) 24

values *see* principles-based approaches

variation: high variation phase 142–146, *144*, *145*; reducing variation phase 146–149, *146*, **148**

verbal fluency 146

video conferencing 215

virtual assessments case study 212–219

visual management boards (VMBs): assessment systems 71, 72, 86, 100; Commonwealth Bank of Australia 188, 193, 194, 195; maturity levels 86; NBN Co. 166; STMicroelectronics 116, *119*, 121–122, 124–126, *127–129*; virtual assessments case study 217, 218, 219

Voice of the Customer (VoC):
  Airbus Australia Pacific 180;
  designing a behavioral assessment
  system 89, 97

wearable technology 213–218
what, the 252
why, the 252
Wilson, Mike 168–169